ACKNOWLEDGEMENTS:

This text is dedicated to the memories of Rose J. Johnsen and Dr. Kevin McCready. Ms. Johnsen was a pioneer in internet home school initiatives and in advocacy for children. Dr. McCready established the first drug free clinics to help individuals restore meaning to their lives and treated each with enormous dignity and respect. This text is also dedicated to my friend and colleague Dr. Clancy D. McKenzie and to all the children I have had the privilege of helping.

INTERNATIONAL CENTER FOR HUMANE PSYCHIATRY

Dunmore, Pennsylvania, USA

www.humanepsychiatry.info

CHAPTER ONE: MEETING THE TRUE NEEDS OF OUR CHILDREN DIAGNOSED AS 'ADHD'

How should one look upon Attention Deficit Hyperactivity Disorder (ADHD) and what is the effective way to aid those who are given this diagnosis? There has been considerable debate as to whether or not ADHD is a genuine disorder. Psychiatrist and professor Robert Hedaya (1996, pg. 140) mentions that an examination by Hartmann in 1993 felt that ADHD is actually normal variant of human behavior that doesn't fit into cultural norms.

In addition, there is no objective test for this disorder. Hedaya (1996, pg. 140) mentions that a commonly used test is the TOVA (test of variables of attention), a test where the client must use a computer and hit a target at various points. This test is designed to measure the person's response time and distractibility. However, Hedaya (1996, pg. 140) notes, this tool cannot be relied upon to make or exclude the diagnosis in and of itself. Hedaya (1996, pg. 268) notes that there has been controversy in the use of stimulants for the treatment of ADHD, he states, medications alone do not provide adequate or full treatment in this disorder.

Hedaya (1996, pg. 269) notes that the most serious risk in the use of methylphenidate (Ritalin) for ADHD is that about 1% of these children will develop tics and or Tourette's Syndrome. Hedaya asks the question," One might wonder-, why use methylphenidate at all?" Hedaya argues that the side effects involved in the use of methylphenidate are mild. However, he notes that side effects include nervousness, increased vulnerability to seizures, insomnia, loss of appetite, headache, stomachache, and irritability. Hedaya (1996, pg. 271) argues that the causation of ADHD lies in problems in dopamine regulation in the brain and states that stimulants work by stimulating dopamine in the brain and thus the symptoms of ADHD are lessened.

However, previously Hedaya states that Zametkin (1995) noted that stimulants have the same effect in both those diagnosed as ADHD and those who are not (Hedaya, 1996, pg. 139). Dr. William Carey of the Children's Hospital of Philadelphia commented at the National Institutes of Mental Health Consensus Conference in 1998 that the behaviors exhibited by those considered ADHD were normal behavioral variations. A Multimodal Treatment Study was conducted by the National Institutes of Mental Health in 1999 in regards to ADHD. Psychiatrist Peter Breggin and the members of the International Center for the Study of Psychiatry and Psychology challenged the outcomes of this study because it was not a placebo controlled double blind study. Breggin also argues that that the analysis conducted of behaviors in the classroom of those children studied showed no significant differences between those children receiving stimulant medications versus those who only were utilizing a behavioral management program (MTA Cooperative Group, 1999a, pg. 1074). Breggin notes that there was no control group in the study of untreated children and that 32% of the children involved in the study were already receiving one or more medications prior to the onset of the study. Of those in the study who were the medication management group, they numbered only 144 of which Breggin finds to be enormously small.

Breggin states that in the ratings of the children themselves that they noted increased anxiety and depression however this was not found to be a significant factor by the investigators. Breggin also believes that the study was flawed in that drug treatment continued for 14 months whereas behavioral management was utilized for a much shorter duration. Breggin argues that the behavioral management strategies, which involved mainly a token economy system, were ineffective as well and did not take into consideration family dynamics but regardless, the study still showed that there was no difference between the populations treated with drugs versus those undergoing behavioral management solely. Breggin notes that many of the children receiving medications had adverse drug reactions, which consisted of depression, irritability, and anxiety. 11.4% reported moderate reactions and 2.9% had severe reactions. However, Breggin also states that those reporting the adverse drug reactions were not properly trained, but were rather only teachers and/or parents.

    The study, as Breggin concludes, showed no improvement in the children treated with medications in the areas of academic performance or social skill development. Breggin feels that the study was improper in that all of the investigators were known to be pro-medication advocates prior to and after the study. Breggin states that Ritalin and other amphetamines have almost identical adverse reactions and have the potential for creating behavioral issues as well as psychosis and mania in some individuals. Breggin argues that these medications often cause the very behaviors they are intended to treat. He notes that children treated with these medications often become robotic and lethargic and that permanent neurological tics can result.

    In his textbook, Attention Deficit Hyperactivity Disorder, Russell Barkley, an advocate for the use of methylphenidate in the treatment of ADHD, notes that there is little improvement in academic performance with the short-term use of psychostimulant medication. Barkley also acknowledges that the stimulant medications can affect growth hormone but at present there is not any knowledge of the long-term effects on the hypothalamic-pituitary growth hormones. Barkley (1995, pg. 122) also states, at present there are no lab tests or measures that are of value in making a diagnosis of ADHD.

Sidney Walker, III, (1998, pg. 25) a late board-certified neuropsychiatrist comments that a large number of children do not respond to Ritalin treatment, or they respond by becoming sick, depressed, or worse. Some children actually become psychotic - the fact that many hyperactive children respond to Ritalin by becoming calmer doesn't mean that the drug is treating a disease. Most people respond to cocaine by becoming more alert and focused, but that doesn't mean they are suffering from a disease treated by cocaine. It is interesting to note Walker's analogy of Ritalin to cocaine. Volkow and his colleagues (1997) observed in their study, (methylphenidate, like cocaine, increases synaptic dopamine by inhibiting dopamine reuptake, it has equivalent reinforcing effects to those of cocaine, and its intravenous administration produces a high similar to that of cocaine. Walker (1998, pg. 14-15) that in addition to emotional struggles of children leading to ADHD-like behavior, that high lead levels, high mercury levels, anemia, manganese toxicity, B-vitamin deficiencies, hyperthyroidism, Tourette's syndrome, temporal lobe seizures, fluctuating blood sugar levels, cardiac conditions, and illicit drug use would all produce behaviors that could appear as what would be considered ADHD however Walker feels that these issues are most often overlooked and the person is considered to be ADHD.

F. Xavier Castellanos states at the 1998 Consensus Conference that those children with ADHD had smaller brain size than those of children who were considered to be normal. However, Castellanos reported as well that 93% of those children considered

ADHD in the study were being treated long term with psychostimulants and stated that the issue of brain atrophy could be related to the use of psychopharmacological agents. Dr. Henry Nasrallah from Ohio State University (1986) found that atrophy occurred in about half of the 24 young adults diagnosed with ADHD since childhood that participated in his study. All of these individuals had been treated with stimulants as children and Nasrallah and colleagues concludes that cortical atrophy may be a long term adverse effect of this treatment. Physician Warren Weinberg and colleagues stated, a large number of biologic studies have been undertaken to characterize ADHD as a disease entity, but results have been inconsistent and not reproducible because the symptoms of ADHD are merely the symptoms of a variety of disorders. The Food and Drug Administration has noted (Walker, 1998, pg. 27) that they acknowledge that as of yet no distinct pathophysiology (for ADHD) has been delineated.

There has been concern as well about the addictive component of psychostimulants. The Drug Enforcement Administration (1995c) reports that it was found that methylphenidate's pharmacological effects are essentially the same as those of amphetamine and methamphetamine and that it shares the same abuse potential as these Schedule II stimulants.

Breggin states that psychiatrist Arthur Green in the Comprehensive Textbook of Psychiatry published in 1989 reported that all commonly diagnosed disorders of childhood can be linked to abuse and/or neglect. Abuse and neglect produces difficulties in school, such as cognitive impairment, particularly in the areas of speech and development, combined with limited attention span and hyperactivity. (Breggin, 1991, pg. 274)

Being that ADHD is a subjective diagnosis and that stimulant treatment has been shown to have risk as detailed above, what is the effective alternative to aiding those who have been diagnosed ADHD and what actually is underlying the difficulties that these individuals may be manifesting? Psychologist and educator Michael Valentine (1988) suggests that it is necessary to "love your children, care about them, do as much as possible to have them grow and develop, teach them social skills, and teach them how to identify and express their feelings and to become uniquely human; but at the same time, care about them and love them enough to give them guidance, structure, limits, and control as they need it.

Valentine advocates a psychosocial approach to aiding children and adolescents who would be considered to be ADHD. Psychiatrist Peter Breggin also advocates this approach and feels that it is necessary for parents to feel empowered and for their to be a compassionate therapeutic adult in the lives of these children. Breggin (1998, pg. 308-310) feels it is necessary to examine the effects of institutionalization and placement on children as well as the effects of psychiatric stigmatization (that is, the effects on esteem of receiving the label of ADHD itself). It is necessary to examine the experience of the child and if they have suffered physical, sexual, or emotional abuse from adults, or have experienced peer abuse. It needs to be examined if they have an appropriate educational setting and if any conflicts exist with instructors or if the educational environment is stressful to them.

Psychiatrist William Glasser (2003, pg. 31-32) comments in this regard, pediatricians are being called in to diagnose schoolchildren who do not cooperate in school because they don't like it as having attention deficit disorder or attention deficit hyperactivity disorder. Treating them with a narcotic drug is only confirming what many psychiatrists and pediatricians already believe: that it's better to use drugs than to try to apply their prestige and clout in the community to the real problem: improving our school s so that students find them enjoyable enough to pay attention and learn in an environment where drugs are not needed. This misguided psychiatric effort has created an epidemic of drug treated mental illness in the schools.

Breggin continues that it is also necessary to examine the environment the child lives in and the stressors around them. It is necessary to build relationship and collaboratively design structure and limits with the child or adolescent (Breggin, 1998, pg. 318) Breggin feels it is necessary to train parents in relationship building with their children and in working through situations of conflict. He states, parent management training has consistently proven successful in improving parent self-esteem, in reducing parent stress, and in ameliorating ADHD-like symptoms, especially negative attitudes toward parental authority and aggression.

David Stein (2001, pg. 236-238) has detailed a drug free approach to aiding children who are diagnosed as ADHD who Stein prefers to call highly misbehaving children. In this program, known as the Caregiver's Skills program, Stein states it is necessary to treat your child as normal and not diseased. He states that the children should not be taking any medications, as they are risky for the child's health and merely blunt behaviors. Stein argues, if the behaviors don't occur, we can't help (them) learn new habits.

The program encourages social reinforcement rather than material reinforcement, encouraging parents to refrain from excessive prompting and coaxing. The program encourages development of target behaviors and consistent encouragement and social reinforcement as well as consistent consequences for misbehavior. The program encourages the self-assessment and evaluation of the child of their own behaviors.

CHAPTER TWO: THINKING OUTSIDE THE BIO-PSYCHIATRIC PARADIGM

As a person working within the mental health profession, I once challenged colleagues to examine the work that they do without using the terms 'mental illness', 'treatment', 'diagnosis' or the like. I could see that initially this was a challenging exercise, however the key terms that came forward were: conflict resolution, mediation, coaching, guiding.

These terms become important when we realize that those who have been labeled as seriously distressed and 'mentally ill' are individuals who have undergone conflict in their lives. These individuals are seeking a voice. Their actions are not random, but rather seek to communicate their experiences. Some individuals who have endured serious trauma begin to speak of their experiences in a metaphorical sense. The role of the therapist should be to help this individual find meaning in this experience, uncover unmet needs, and to listen and be able to understand this experience.

Biological determinism and the theory of distress arising from so-called chemical imbalances is a popular and majority idea in the mental health field today. However, there is no evidence to support such a concept. Such an idea helps to further the profits of the pharmaceutical industry who are able to make lifelong mental patients in need of their products through the promulgation of such chemical imbalance concepts. The President of the American Psychiatric Association recently stated that there is no 'clear cut test" to demonstrate chemical imbalances. Dr. Eliot Valentstein goes further to state, "Elliot Valenstein, Ph.D. says, "[T]here are no tests available for assessing the chemical status of a living person's brain." The late Dr. Loren Mosher who had headed Schizophrenia research for the National Institutes of Mental Health stated, "...there are no external validating criteria for psychiatric diagnoses."

Psychiatry has an oppressive history. In the early 1900's the theory of the day was the concept of improper blood flow to the brain causing mental illness which led to the invention of oppressive treatments such as the 'swivel chair' which was designed to redistribute blood flow to the brain. In Nazi Germany, psychiatry was used as a tool of social control as well as in the Soviet Union where political dissenters were labeled mentally ill and institutionalized. The leader of the Bosnian Serb militia was Radovan Karadzic, a psychiatrist. Psychiatry has been used to oppress gay and lesbian persons who until 1973 were designated 'mentally ill' and African-Americans have also been oppressed as early psychiatry considered them to be suffering from mental illness known as "negritude" and had lower intelligence than that of Caucasian persons. Journalist Robert Whitaker in his text, Mad in America details the oppression that continues to this day in mental institutions in our own country.

Today's mental health profession has been commandeered by biological psychiatry. Biological psychiatry has sought to look upon the behaviors and emotional world of our children and adolescents as the result of 'broken brains' and views children expressing distress as somehow disordered. Many children today who show any type of inappropriate behaviors are often immediately being labeled as ADHD and being prescribed stimulant medications such as Ritalin, Adderall, or Dexedrine among others. There is no test for ADHD and neurological testing shows these children to be perfectly normal. Dr. William Carey of Children's Hospital in Philadelphia states, "common assumptions about ADHD include that it is clearly distinguishable from normal behavior, constitutes a neurodevelopmental (brain) disability, is relatively uninfluenced by the environment (home, school)...all of these assumptions...must be challenged because of the lack of empirical support and the strength of contrary evidence...what is now described in the US as ADHD is a set of normal behavioral variations. This discrepancy leaves the validity (of ADHD) in doubt. The U.S. National Institutes of Health Consensus Development Conference on ADHD in 1998 reported, " we have do not have an independent, valid test for ADHD, and there are no data to indicate that ADHD is due to a brain malfunction...and finally, after years of clinical research and experience with ADHD, our knowledge about the cause or causes of ADHD remains speculative." Further, Dr. Edward C. Hamlyn, a founding member of the Royal College of General Practitioners in 1998 stated, "ADHD is fraud intended to justify starting children on a life of drug addiction." The U.S. Surgeon General Report declares, "the exact etiology of ADHD is unknown." Lastly, Dr. Joe Kosterich, Federal Chair of the Australian Medical Association states, " "The diagnosis of ADD is entirely subjective.... There is no test. It is just down to interpretation. Maybe a child blurts out in class or doesn't sit still. The lines between an ADD sufferer and a healthy exuberant kid can be very blurred." A report completed by the Oregon State University Drug Effectiveness Review Project (2005) studied 2,287 studies and concluded the following: "No evidence on long-term safety of drugs used to treat ADHD in young children" or adolescents." Good quality evidence ... is lacking" that ADHD drugs improve "global academic performance, consequences of risky behaviors, social achievements" and other measures. Safety evidence is of "poor quality," including research into the possibility that some ADHD drugs could stunt growth, one of the greatest concerns of parents. Evidence that ADHD drugs help adults "is not compelling," nor is evidence that one drug "is more tolerable than another." The way the drugs work is, in most cases, not well understood."

What we are experiencing are children in conflict. We cannot blame and denigrate the child and not respect his dignity. We cannot label and suppress behaviors. If a child is conflict, we must take responsibility to see why this child is in conflict and to use responsible and carefully planned interventions to aid this child in being successful. Some believe that they see enormous benefits from children on medication. I will give them this benefit, only if we see suppression of behavior, basically chemical restraint, as our concepts of 'what works', of 'progress' or 'success'. Because something 'works' does not imply it is good or ethical. I could probably scientifically validate that strapping a child to a chair would also aid with supposed hyperactivity, but this would be aversive and illegal. But once again, we are taking normal children and drugging them to suppress their behaviors. In order to achieve such a 'result', just what is going on in this child's body? Stanley I. Greenspan, a clinical professor of psychiatry at George Washington University states that, " The growing use of medication on their own is a worrisome trend while more and more people on Prozac or Ritalin are becoming bolder and less distractible, at the same time, more and more people are altering their moods without understanding what is happening to them or how it relates to their core personalities." He also states, "given appropriate nurturing, many affected children may not require medication." University of South Florida Professor of Psychology, Diane McGuiness comments, "The first factor of being put on drugs is to attribute your bad behavior to factors beyond your control. Drugs become a substitute for learning self-discipline. This problem is compounded when children are taken off medication and problem behavior initially rebounds to fantastic proportions. Second, longitudinal studies have confirmed that children on drugs actually deteriorate in academic performance over time. And we must consider the sense of worthlessness most of these young people experience.(McGuiness, 1985). Paul Wender, M.D. lists criteria when beginning medication, he states that a child must first understand why he is receiving medication, yet as Greenspan states above, this isn't always happening. Wender states, "Most acknowledge problems in his own behavior that he himself does not like, so that -he will not feel that medicine is being given to him simply so that other people can tolerate him more." Now, Wender is one who began the first tests on the use of methylphenidate and is in support of its use in treatment. Greenspan comments, "working with the strengths of a child can create motivation." A child needs to be able to recognize and be motivated to change behaviors and work on strengths. Even Wender states that getting a child to 'label' behaviors is effective, that a child must recognize what is appropriate and what is not, and that parents should not encourage the idea that because the medication was wearing off or so forth that such excuses a level of knowledge and responsibility for certain behaviors. My concern lies too in that whereas some may feel medication to create some responsive in level of focus and so forth, it comes with a cost in side effects. Some may take the view that the potential for progress outweighs the potential side effects. This is where I disagree, and feel it better to avoid that which would cause any side effects, that psychotherapy alone can manage the difficulties. These are some of the things that bring alarm to me. Wender states, "Most common side effects of the stimulant medications are appetite loss...difficulty in falling asleep." He suggests the use of a small dose of sedative 'major tranquilizer' an hour before bedtime to solve this in some cases. So, here a see a cycle of drugs needing to be used and that's worrisome. Wender states, "Research is being conducted to determine the -exact- effects of stimulant medication on growth." This tells me they are prescribing something, which they really do not know yet what the effects are on growth. He states as well that stimulant medication IS addictive in adults, but says, "The results suggest that there is no increased risk for drug abuse associated with treatment, although -more research is needed to rule this out conclusively.-" Here again, if it is addictive to adults, I ask, why not children too, and he says that research is yet conclusive. With effects on the cardiovascular system, Barkley states, "studies have -not- specifically addressed this important issue." So, they are prescribing something for which they are unsure of the effects on growth as well as the cardiovascular system. Barkley states as well, "The side effect that should receive serious attention from clinicians is the possible increase in motor or vocal tics produced by stimulant medication." He continues- "It still seems prudent to screen children with ADHD adequately for a personal or family history of tics or Tourette's Syndrome." I recall having a session with a child with the mother first and being informed that he was being treated with Ritalin, I later had subsequent time with the father, and he had visible tics, this really alarmed me in reading about the issue of tics and Tourette's and I had to question not to the family but within myself if this was really the best option for the child faced with this risk. Barkley also states, "Isolated cases may arise in which parents note that a child is no longer 'spontaneous' or childlike in his or her behavior and appears -controlled- or -socially aloof-. This is concerning, and it appears that stimulants do have an affect in gaining control and conformity. Many of the stated results of the stimulant medication are too subjective, and Barkley states clearly that, "an improved ability to master increasingly difficult or higher-level academic material, such as that assessed in achievement tests, has -not- been demonstrated. Here we have an example that it would not have been as a result of stimulant medication if we see academic progress. Barkley also says that 'low and moderate doses of methylphenidate do reduce the frequency of aggression and noncompliance in groups of children but have no appreciate effect on either direction on prosocial or nonsocial behaviors." So, I will acknowledge that stimulants can help with short-term behavioral inhibition, but what about long term? This is my main concern, with the side effects and without evidence of a long-term result and without knowledge of long-term results on growth and cardiovascular development, is this really the best option? Barkley states, " Few studies employing rigorous methodology have evaluated the long term efficacy of stimulant medication. Those that have examined the issue have generally found little advantage of medication over no medication when evaluated over extended periods (Pelham, 1985, Weiss &Hechtman, 1993) Children who had been on drugs but were off at the time of follow-up were not found to differ in any important respect from those who had never received pharmacotherapy." Another concern is the effects in mood, I worked with one child who was already experiencing social withdrawal and was going through the trauma of losing a loved one. After receiving stimulant medication, this intensified. Wender states, "Instead of becoming high or excited, these drugs in general calm down ADHD children and sometimes they may even become somewhat sad." Barkley states, "some children may evidence various mild negative moods or emotions in reaction to stimulants...Some children describe feeling 'funny', 'different' or dizzy as a function of medication." What about self-esteem and confidence, Greenspan acknowledges that creativity can be affected, and Barkley states, " some concern has been raised that diminished self-esteem could be a emanative effect of methylphenidate as children may attribute the source of their success while on medication to external rather than internal factors."

The Eli Lilly company has been marketing a new drug for those who are labeled as ADHD known as Strattera. This drug is purported to be a non-stimulant medication, however the side effects are similar. Rather than effecting the dopamine system as do the stimulants such as methylphenidate and dextroamphetamine, it works upon the norepinephrine neurotransmitter. Strattera is considered a Norepinephrine reuptake inhibitor. Norepinephrine is the brain's adrenalin. Norephinephrine is involved in the increased rate and force of the heart muscle, constriction of heart muscles, pulmonary function (Hedaya, 1999). If these functions are increased, it would be evident that Strattera could produce possible untoward effects on the cardiovascular system. It is interesting to note that in the safety information that Eli Lilly provides on its website, it refers to possible hazards to those who have heart disease or high blood pressure. Information provided by Eli Lilly accompanying prescriptions of this drug note the possibility of tachycardia, and increased blood pressure. Tourette's disorder, though the etiology is not fully known is being examined as difficulties arising in the metabolism of dopamine, serotonin, and norepinephrine. it is known that stimulant drugs can produce Tourette's like behavior in some children (Breggin, 1998). If Strattera affects the norepinephrine system, then it would seem evident that the possibility of such Tourette's like behavior could also become manifest in some children using Strattera. Therefore, though Strattera is being marketed in the fashion of being a non-stimulant drug, its ill effects are quite similar to that manifested by the stimulant medications. Eli Lilly's website notes that growth suppression is a common side effect and needs to be monitored in children making use of this drug. Loss of appetite and weight loss is also seen. The most common side effects as listed by Lilly are upset stomach, decreased appetite, nausea and vomiting, dizziness, tiredness, and mood swings. These are not unlike that associated with the stimulant medications. Lilly states in its press release in regards to Strattera's introduction: "It's not known precisely how Strattera reduces ADHD symptoms. Scientists believe it works by blocking or slowing reabsorption of norepinephrine, a brain chemical considered important in regulating attention, impulsivity and activity levels. This keeps more norepinephrine at work in the tiny spaces between neurons in the brain." If we examine this statement carefully, we see it states 'it is not precisely known', therefore once again a drug is being prescribed whose effects are not fully known for a'disorder' whose psychopathology is not yet delineated. Clinical trials for Strattera have been limited and any information on long term effects has only been studied by Lilly itself. It is interesting to note that before Strattera was actually placed on the market and had just received FDA approval that the stocks for Eli Lilly rose 6% at the announcement (CBS Marketwatch, November 27, 2002). Lilly is aware that it will profit highly by being able to market a drug as a non-stimulant (though its ill effects are similar), that is not a Schedule II drug thus less subject to scrutiny and regulation. Hemant K. Shah, an independent analyst quoted in an AP Health News Report (August 15, 2002) states that Strattera's market potential is large at a time when Eli Lilly is seeking to offset recent setbacks. , "Parents who have refused stimulant dangers because of their knowledge of the hazards involved will now be coerced to utilize Strattera being led to believe it is somehow safer because it does not fall into the category of a stimulant/ Schedule II drug.

Dr. Howard Gardner (1983) identified the concept of multiple intelligences. Gardner identified the following: Linguistic intelligence ("word smart"), Logical-mathematical intelligence ("number/reasoning smart"), Spatial intelligence ("picture smart"), Bodily-Kinesthetic intelligence ("body smart"), Musical intelligence ("music smart"),Interpersonal intelligence ("people smart"), Intrapersonal intelligence ("self smart"),Naturalist intelligence ("nature smart"). Gardner identified that most educational settings are geared towards the child who utilizes linguistic or logical-mathematical methods of learning. Therefore, those who utilize other learning styles and intelligences fall outside the mainstream educational system and thus are prone to be considered 'learning disabled' or to have "ADHD". Capturing children's interest in the learning process by adapting to their learning styles through creating a zeal for learning through such use of art, music exercises, role plays, reflection, and other means could provide an effective solution towards meeting the needs of our children.

Glasser has designed the concept of quality schools where children learn both independently and collaboratively, where mutual respect is a key concept and the method of instruction and management is completed through the use of choice theory rather than external control and punishing. Children are given a sense of responsibility and involvement. Critical thinking is encouraged and educators monitor themselves through student and team feedback. Student's learning is based on application and usefulness rather than rote learning. The building of meaningful, sustainable relationships are of utmost importance. Students are encouraged in demonstrating competency rather than rehashing material.

David Healy, former secretary of the British Psychopharmacological Society notes that vast influence of the pharmaceutical industry in skewing clinical trials. Recently, the FDA has issues serious warnings on a number of SSRI anti-depressant drugs in regards to increased suicidality in children and adolescents. Healy has argued that the pharmaceutical companies, such as Eli Lilly were very aware of the complications that these drugs caused but withheld information from being published. In May 25th, 1984, a communication from Eli Lilly US from Lilly Bad Homburg stated the following: "Considering the benefit and the risk, we think this preparation totally unsuitable for the treatment of depression.' Lilly has recently settled a $690 million lawsuit in regards to increased diabetes risk with the use of its drug Zyprexa.

It is interesting to note that nearly all of the children involved in the situations of school shootings were involved in the mental health system and undergoing treatment with one or more psychotropic agents: Shawn Cooper, a 15-year-old sophomore at Notus Junior-Senior High School in Notus, Idaho, fired a shotgun at his fellow students in April. Cooper was on Ritalin. Thomas Solomon, a 15-year-old at Heritage High School in Conyers, Georgia, shot and wounded six classmates in May. Solomon was on Ritalin. Kip Kinkel, a 15-year-old at Thurston High School in Springfield, Oregon, killed his parents and two classmates and wounded 22 other students last year. Kinkel was on Ritalin and Prozac, an anti-depressant. Eric Harris, one of the Columbine High School killers, was on the anti-depressant drug Luvox. Rod Matthews, 14, beat a classmate to death with a baseball bat in 1986 in Canton, Massachusetts. Matthews had been on Ritalin since the third grade. Yale researchers, as published in the March 1991 Journal of the American Academy of Child and Adolescent Psychology, found in their study of

Prozac at least one 12-year-old who started having nightmares. The boy dreamed of killing his classmates at school until he himself was shot. The researchers took the boy off Prozac and he recovered. Then they put him back on the drug, apparently thinking that the anti-depressant could not have caused the nightmares. Once drugged again, the boy started to have acute suicidal thoughts and tendencies. Psychiatrist Peter Breggin has noted that a number of the SSRI's can be associated with triggering manic episodes and violence. In one clinical trial involving stimulants the rate of psychosis as an adverse drug reaction was 4%. This number may at first appear small, but when we examine that nearly 6,000,000 children are receiving stimulants in the US, this tells us that the possibility of 240,000 episodes of psychotic reaction could occur. In addition, as extended release drugs are being utilized, this makes the potential for adverse drug reactions more serious as the drug stays within the system for a longer period.

Two organizations that claim to be support groups for families can actually be considered front groups for the pharmaceutical industry. One such group is CHADD and the other NAMI. CHADD's events have been frequently sponsored by grants from Ciba Pharmaceuticals who makes Ritalin and in the first year, the majority of the funding for this non-profit organization came directly from the pharmaceutical companies including grants from Ciba, Pfizer, and Alza Pharmaceuticals among others. NAMI also receives large grants from the pharmaceutical industry and promotes the use of psychotropic drugs and the theory of 'chemical imbalances."

How do we deal with our seriously distressed children and adolescents? Adolescents are in a period of seeking autonomy and self-determination. These qualities can aid them in becoming agents of active transformation in their own lives. For one to recover from distress they are in need of being able to regain hope and to have an effective exercise of their free will. (Breggin, 1996). Adolescents based on their experiences formulate thoughts and feelings and begin to create values and meanings for themselves. Those adolescents who are suffering from serious emotional distress have become lost on this path to finding meaning in their lives. Once this occurs, they begin to develop anguish and self-defeating responses to life. This creates in them anxiety and despair leading towards what some would call 'madness' (Breggin, 1991). These adolescents must learn to feel empowered once again, and not to feel labeled as an 'it', not to be viewed through the lens of their particular diagnosis and categorization they have been ascribed. These adolescents need coaches and individuals who will aid them compassionately and empathetically in navigating and negotiating through life's stresses. The therapist and others must look upon the distressed adolescent with dignity. To look upon the adolescent through 'scientific' or 'objective' means leads us to the tendency to diagnosis and control the person, to impose our own abstract and potentially oppressive category upon them and to manipulate the outcome. Physical interventions, such as psychotropic drugs, restraints, and enforced confinement to mental hospitals or residential treatment facilities are a part of this desire to control rather than truly aid and come to an understanding of the distress the adolescent is experiencing (Breggin and Breggin, 1993, a&b). Psychotropic medications with these seriously distressed individuals only deal with symptoms, they blunt certain functions to make the person more tolerable and amenable to societal expectations. Psychotherapy, on the other hand, focuses on the subjective changes in patient's feelings and on actual changes in lifestyle or conduct of life (Fisher & Greenberg, 1989). Based on the viewpoints of biopsychiatry, adolescents who are medicated and placed in mental hospitals are labeled as improved when they conform to hospital demands or receive discharge. However, what is not examined is, how do the patients themselves actually feel? An estimated 180,000 to 300,000 young people a year are placed in private psychiatric facilities. These children and adolescents often feel powerless in these placements. But as mentioned above, it is the need for feelings of empowerment and hope that will lead to a genuine recovery from distress. Psychologist D.L. Rosenhan lead a study where 'pseudopatients' had themselves admitted to psychiatric hospitals to experience them first hand and report on this experience. Rosenhan reported in an article appearing in the January 19, 1973 issue of Science, "Powerlessness was evident everywhere…He is shorn of credibility by virtue of his psychiatric label. His freedom of movement is restricted. He cannot initiate contact with staff, but may only respond to overtures as they make. Personal privacy is minimal…" With children and adolescents it is easier to rationalize away their rights and control becomes more arbitrary and complete (Breggin, 1991). Psychiatrist Peter Breggin states that in such an environment 'it is hard for a child to resist feeling spiritually crushed, abandoned, and worthless under such conditions. With a less formed sense of self than an adult has, a child is less able to resist the shame attached to being diagnosed and labeled a 'mental patient'. Children may also find it much harder to conform to institutional life. They are naturally energetic, rambunctious, at times strident, often noisy, and resistant to control. If a boy doesn't conform, he is considered 'ill' and can be subjected to physical restraints, solitary confinement, and toxic drugs. (Breggin, 1991). It should be mentioned that the drugs commonly used for severely distressed adolescents are the same as those used for adults, most frequently the neuroleptics. These medications are reported to cause lack of energy, painful emotions, motor impairment, cognitive dysfunction and tend to 'blunt; the personality of the treated patients as well as having a risk for the development of tardive dyskinesia, a permanent and debilitating neurological problem (Gualteri and Barnhill, 1988). These drugs subdue the adolescent into conformity by blunting the brain, but never do they teach the child how to develop meaning, how to cope, nor do they allow the adolescent to express his pain and emotional distress that is within. The adolescent is merely sedated to make his behaviors more manageable to adults. The adolescent learns nothing. The adolescents who are suffering from severe emotional distress are in conflict. They have internalized feelings of guilt, shame, anger, anxiety, and numbing. These adolescents instead of coercive and intrusive 'treatments' need the ability to find a safe place where coercive power is replaced by reason, love, and mutual attempts to satisfy their basic needs. These adolescents because of their distress have broken away from the accepted realities, they have sought to recreate their existence, for some a more primitive existence (Schilder, 1952). The feelings of anxiety that an adolescent may experience are linked to a fear of being and belonging (Stern, 1996, pg. 12) Depression, mania, and anxiety are all linked together and are indicative of trauma. The adolescent being a shattered person seeks an escape by altered perception. We must begin to realize that all behaviors and experiences have meaning, even those things that may appear the most 'odd' to us. The symptoms labeled to be schizophrenic exhibited by certain adolescents in distress 'may be understood as manifestations of chronic terror or defense against the terror (Karon, 1996). This is often expressed as anger, loneliness, and humiliation. The therapist and others must convey to the adolescent that he wants to understand, that the client is helpable, but it will take hard work (Karon, 1996). The therapist must forge an alliance with the adolescent, aiding them to understand the real dangers and to be able to develop appropriate coping mechanisms. These adolescents are often viewed as dangerous themselves but the majority are not. They need to be hard, and forging this alliance will give them the needed voice leading to their recovery. Hallucinations that are experienced by the seriously distressed adolescent are actually repressed thoughts and feelings coming outward, the unconscious into the conscious. Delusions are the adolescent transferring experiences from their past without having the awareness that it is past (Karon, 1996, pg. 36). The therapist can guide in interpreting the meaning of these hallucinations and delusions and once the adolescent is gently approached with their underlying meaning, these events can dissipate. Delusions are also connected with an attempt to find a systematic explanation of our world, to find meaning. A person who has experienced severe distress has lost this meaning and thus develops unusual ways of seeking to make sense of their experiences and the world around them (Karon, 1996, pg. 38). The therapist can gently call the adolescent's attention to inconsistencies but at the same time respect their vision. The results of a psychosocial approach to those with severe emotional distress has been proven to be more effective than the current biopsychiatric methods as evidenced by a study by Loren Mosher, MD where he took schizophrenic adults who were on either very low doses or no medication, and offered them a 'safe place' with non professional staff residing with them and sharing in their daily experiences. A 2 year follow up of these patients noted higher levels of success and progress than their counterparts who were subjected to neuroleptics and psychiatric hospitalization (Mosher, 1996, pg. 53) The model known as the Soteria project was based on principles of growth, development, and learning. All facets of the distressed person's experience were treated by the staff as 'real' (Mosher, 1996, pg. 49) Limits were set and mutual agreements made with the patients if they presented as a danger to themselves or others. Such a model could be adapted to use with adolescents, offering them the need for compassion, empathy, and finding that 'safe' place, restoring within themselves a feeling of worth and dignity, that will lead to their ability to address the issues of their distress and traverse towards recovery.

In today's mental health system there is a pattern of fraud and coercion that takes way the freedoms and dignity of children and their families. Children are receiving stigmatizing labels and being prescribed psychotropic drugs with many untoward effects. Psychiatrist Thomas Szasz, MD made the comment that if an individual hit us with a blackjack and robbed us of our dignity we would call them thugs, yet psychiatrists label and drug children and rob them of their dignity and nothing is said. All in the name of profit. Rarely, if never are the families given informed consent. Szasz has also stated, "From a sociological point of view, psychiatry is a secular institution to regulate domestic relations. From my point of view, it is child abuse." Families are provided with literature that appears so matter of fact but is funded by the pharmaceutical companies and tainted with their bias. According to the Poughkeepsie Journal, the 'support' or should it be said front group for Children diagnosed with Attention Deficit Hyperactivity Disorder received substantial funds from the pharmaceutical companies: "CHADD received $315,000 from drug companies in the year ending June 2000, about 12 percent of its budget."

Children are being beaten, improperly restrained, physically and sexually abused, and emotionally scarred in residential treatment programs. Juvenile probation officials are failing to understand the emotional distress of our children, they are submitting to this "psychiatric Gestapo". Educators rather than finding new methods of shaping our children's learning are falling into the trap of psychiatric 'solutions' as well. Never could it be that a school has simply failed to help a child learn, rather it is always the child denigrated and labeled as 'disordered'. There are loving and concerned parents, and there are others who lack love and compassion towards their children. There are loving and concerned parents who become duped by the 'professionals'.

Some of tried to utilize the analogy of real, demonstrable physical diseases such as diabetes to compare with ADHD. As prior mentioned, there is no demonstratable physical abnormality with ADHD, thus it is not a disease. Insulin is a naturally occurring substance in the body by which one with diabetes either produces too much or too little. Who has ever heard of a person having a Ritalin deficiency? In addition, if I were to walk in to my physician and he tests my heart rate as normal, the nurse enters and startles me, and my heart rate is measured again, and my physician thus says I have heart disease, none would probably ever visit this physician again. But this is what is happening with psychiatrists. They fail to recognize the experience behind behaviors, and just seeing the behavior conclude a so-called disease is present.

I share this scenario because sadly it is becoming a frightening reality: A child is considered overly active and has behavioral issues at school. The school staff may recommend psychiatric intervention and even go as far as to say that medication is necessary, even designating which one. This now illegal after the passage of the Child

Medication Safety Act, however it has and still continues to occur. The child sees the psychiatrist for a brief session- it is not thoroughly examined if the child has any physical conditions, allergies, etc. Immediately the child is labeled and given a dose of psychostimulant. The child develops side effects such as weight loss, insomnia, and possible tics. In order to counteract the insomnia, a new drug such as Klonidine is added. The child develops emotional lability and has crying episodes and manic behaviors. The psychiatrist is seen again for a brief time, and on this visit its determined that 'bipolar is emerging'. The child is then given Depakote or some other mood stabilizer. The child now must receive regular blood tests to insure that liver toxicity does not arise. The child is not overly active, he is quite docile, so it is reported that improvement has occurred. However, with the combination of drugs, he develops some psychotic like symptoms where he feels something is crawling on him and has some hallucinations. The psychiatrist is consulted again, and its determined that bipolar with psychotic features exists or maybe even the possibility of childhood schizophrenia. The child is then given Risperdal or another neuroleptic. Strangely, the child begins developing unusual jaw movements and muscle rigidity. The parents are concerned and ask the psychiatrist if this is medication related and if the child is overmedicated. The psychiatrist brushes off the question and prescribes Cogentin (used for Parkinson's) to alleviate the neurological problems caused by the development of tardive dyskinesia but fails to remove the offending agent. The child's behavior becomes more unusual and bizarre leading to hospitalization where medications are raised and adjusted and new ones added. Then the recommendation comes from the psychiatrist that it would be better for the child to be moved to a residential treatment facility. While in the residential facility, the child is frequently restrained and is injured, he is placed with other children with serious emotional and behavioral distress. He is discharged home having absorbed a lot of new negative behaviors from peers, lacking knowledge of the outside world, and with few skills. So, once the child nears adulthood, it is recommended that he live in a group home where he can be cared for and the psychiatric regiment can be maintained. The child has been 'treated.'

What is the solution to resolving the insanity of the mental health system? First, we must stop looking through the eyes of a medical model, where we see children as broken and disordered and attempts are made to attributing their behaviors and emotions solely to a malfunctioning brain. There is no evidence supporting the psychopathology of a number of disorders. The linkage between the pharmaceutical companies and psychiatry needs to be evaluated as well as the information that is disseminated via the research and materials provided by pharmaceutical company money. One such example is CHADD, the 'support' group for parents of children diagnosed with ADHD that has received a great deal of his funding from the pharmaceutical companies. The goal should be to examine the underlying factors of a child's behavior, looking at the child with dignity and respect, and seeing the child as one in conflict rather than a person who is disordered. Such stigmatization remains indefinitely, and labels can often become a self fulfilling prophecy and will follow our children for years to come and shape the way that they view themselves and also the way others view them, particularly the educational system. We cannot look to solely the most cost effective solution when our children's lives are at stake. Indeed, providing a prescription may control aspects of behavior and be though to have a 'therapeutic effect' but never gets to the root cause, and whereas it is far less expensive to medicate than to provide ongoing psychotherapy, it is appropriate and compassionate counsel that will make the difference. Second, the realm of psychotherapy must return to its original roots. The word psychotherapy literally means the healing of the soul. We must return the soul to therapy, encouraging therapists to instill within themselves the principles of compassion and empathy that are crucial for any therapeutic relationship to blossom forth. Therapists need to be compassionate and creative, and willing to give additional time and effort to see that a child's needs are met and to also provide community linkages and ongoing support within their environment and to encourage the least restrictive setting for our children. We need to hear our children's voices, even if they speak to us in metaphorical means. We need to listen to the behavior of children as possibly their only way to communicate to us their situations of distress and the impact of living in a disordered world. The coercion of parents and families into forced 'treatments' needs to be eliminated. Third, the educational system must be willing to accommodate to meet the various learning styles of children and not seek to place them in a box of rote learning or limit them to one particular style. Some children may falter in a visual setting and need a hands on approach, whereas others may need other methods of encouraging their effective learning. We must return time, attention, and individuality to the classroom. Fourth, parents need to continue to take an active role in the lives of their children, providing ongoing guidance, validating emotions and not taking a dismissive, disapproving, or hands off approach. Rather, parents must be involved in helping the children develop their own sense of being, and being able to assess themselves. Parents need to avoid nagging their children and becoming entrapped in the propaganda that their children are disordered and need drugs to function. Fifth, our society must change in it attitudes. If we worked towards ending poverty and alleviating social injustice, the rates of emotional distress would definitely decline. We are a society where we try to find our answers to ailments within a simple pill. We are a society that has unfortunately lost sight for the welfare of our children. We are a society where we are prosperous, yet greed often blinds us. Such disorders such as ADHD can be looked upon as a social construct. 90% of Ritalin sales are in the US. This tells us that there is something to be examined within our society that needs correction. Somewhere along the line we have failed our children. This is not to lay blame on any particular individual but to understand that our children are in crisis, and it is up to all of us to take the steps for change. We need to rely less on psychiatry and its devices to solve our problems and more on what we can do within ourselves- to take a holistic approach, to understand the child as a whole person- physical, emotional, and spiritual, and to examine in each of these areas where there may be difficulties that can be alleviated. We need to rely less on others dictating the course of our own and our children's lives and develop workable plan within our own family structure. Nothing will ever be perfect, but even in the most serious disturbances, love and compassion can heal much. We must realize that in some situations within society and within our own lives, we may never be able to evoke complete change. This is the cause of much distress, not problems themselves but how we respond to them. To battle those things beyond our control can lead us to emotional distress, but if we seek live as principled individuals, we can make a difference. What we model to our children and to others has a lasting impact.

    I recall in my beginnings as a mental health professional, I was presented with a child who had suffered neglect and was witness to domestic violence. The child's behaviors were at times aggressive verbally and physically towards the mother and his sibling. The mother took up the suggestion from a psychiatrist to try medications and this child was placed on Zoloft and Adderall. Within a matter of days, emotional lability set in and the child began having suicidal ideations, became increasingly more aggressive, did not sleep, and had frequent bouts of crying. When the medication was weaned off, these events were not present. This child was 6 years old.

Another young person I had met was diagnosed with a psychotic disorder. He had become involved in the juvenile justice system who decided to send him to a residential treatment facility. The young man contacted his mother and asked for me to come and visit him that something 'really bad' had occurred. The incident was said to have occurred on the prior Sunday and I was not able to make my visit until Wednesday. On that day, I met the young man who had choke marks around his neck, was bruised on his shoulders, and had a few cuts. I asked the young man what had occurred and he indicated that a staff had restrained him. These wounds were observed days later, so one can only imagine the force used during the actual restraint and how improperly it was conducted. I made a report to the Child Protective Services. No one at this residential facility would speak to me about the matter and I was later to learn that the staff responsible was terminated. However, Child Protective Services did not file charges or find the abuse to be founded because in their own words, "the child did not suffer enough pain" to warrant such action.

A 9 year old boy became my client after an extensive history of psychiatric hospitalizations which led to a cocktail of psychotropic drugs including anti-psychotics. When I first met the child, I noticed unusual jaw movements and twitching. Concerned about the possibility of tardive dyskinesia, I alerted the psychiatrist involved with the child. The psychiatrist sought to reassure me that tardive dyskinesia is found mainly in the elderly and argued that this child was not experiencing tardive dyskinesia. I remained unswayed in my opinion. Six months later this very psychiatrist diagnosed tardive dyskinesia and recommended Cogentin, a Parkinson's drug to alleviate the spasms. The child is now permanently disfigured.

A 5 year old child prior to coming to me as a client had an extensive history of sexual abuse perpetrated by a relative. The child began acting out in unusual ways in reaction to his traumatic experience. The mother dismissed the allegations of sexual abuse. It took numerous reports to Child Protective Services before any action was taken in regards to the abuse. In addition, the mother had taken the child to a psychiatrist who prescribed Risperdal. The child took a seizure at school, collapsed and was drooling. The child was transported to the emergency room. The medical report which I obtained identified Risperdal as the cause for this occurrence.

I saw a child for an evaluation for wraparound services. The child entered the room barely able to stand, eyes opening and closing, and falling over in his seat. He appeared in an intoxicated state. I was to learn that the psychiatrist had this 12 year old boy on 6 different psychotropic medications.

In my work, I have always sought to find alternatives to the use of these toxic, brain disabling drugs, and for those already taking them to find means to reduce the dependency and look towards eventual withdrawal in consult with medical professionals. In my doctoral dissertation, "Examining the Efficacy of Wraparound Programs and the Efficacy of the Caregiver's Skills Program versus Medication as an adjunct to Treatment" I found that children diagnosed with 'disruptive behaviors disorders' actually faired better using a psycho-social approach than psychotropic drugs. As I have noted many times, drugs never teach new skills, people do. It is time that we re-establish the needed relationships with our children, and invest our time, energy, and our hearts into their lives. It is time to abandon the lie and fraud of so-called 'chemical imbalances' and identify the real root of why are children are in distress.

CHAPTER THREE: THE RELIGION OF BIO-PSYCHIATRY

If we examine the term 'psychotherapy' we will see that it literally means the 'healing of the soul'. Much of emotional distress is actually a spiritual problem, where an individual has lost meaning and hope. In today's society, we have lost sight of spiritual values and many have in turn decided to turn to the religion of bio-psychiatry.

Bio-psychiatry has a creed, the creed is that all problems of life are the result of so-called chemical imbalances. Any professional or individual challenging such conception is branded a heretic and subject to sanction.

The psychiatrist is the priest, and in some cases is also looked upon as a messianic figure. As the priest, the psychiatrist defines the Social Law and the infractions thereof.

The religion of bio-psychiatry has its 'sacraments.' The sacrament of initiation occurs at intake, and confirmation occurs when the individual is indoctrinated into the credo of bio-psychiatry and led to believe that their problems lie in 'wiring' problems in their brains.

The religion of psychiatry has absolution. Through their ritual, adults and institutions are able to be completely absolved from taking any responsibility for meeting the real emotional needs of a person or examining how their own actions and choices could have impacted the challenges faced by the individual.

Ordination occurs at the conferring of the Medical Degree and the completion of a residency in psychiatry.

The religion of bio-psychiatry has the 'sacrament' of Matrimony. Bio-Psychiatry is wedded to the pharmaceutical industry, and what a generous dowry the pharmaceutical industry has provided!

The American Psychiatric Association serves as the "Ecumenical Council" to promulgate the creed of bio-psychiatry and to institute the definitions of the Social Law and the infractions thereof.

The Bible of bio-psychiatry is the DSM-IV.

The Deity is the enormous ego of the bio-psychiatrists themselves.

CHAPTER FOUR: EXAMINING STRATERRA

The Eli Lilly company has been marketing a new drug for those who are labeled as ADHD known as Strattera. This drug is purported to be a non-stimulant medication, however the side effects are similar. Rather than effecting the dopamine system as do the stimulants such as methylphenidate and dextroamphetamine, it works upon the norepinephrine neurotransmitter. Strattera is considered a Norepinephrine reuptake inhibitor. Norepinephrine is the brain's adrenalin. Norephinephrine is involved in the increased rate and force of the heart muscle, constriction of heart muscles, pulmonary function (Hedaya, 1999).

If these functions are increased, it would be evident that Strattera could produce possible untoward effects on the cardiovascular system. It is interesting to note that in the safety information that Eli Lilly provides on its website, it refers to possible hazards to those who have heart disease or high blood pressure. Information provided by Eli Lilly accompanying prescriptions of this drug note the possibility of tachycardia, and increased blood pressure. Tourette's disorder, though the etiology is not fully known is being examined as difficulties arising in the metabolism of dopamine, serotonin, and norepinephrine. it is known that stimulant drugs can produce Tourette's like behavior in some children (Breggin, 1998).

If Strattera affects the norepinephrine system, then it would seem evident that the possibility of such Tourette's like behavior could also become manifest in some children using Strattera. Therefore, though Strattera is being marketed in the fashion of being a non-stimulant drug, its ill effects are quite similar to that manifested by the stimulant medications. Eli Lilly's website notes that growth suppression is a common side effect and needs to be monitored in children making use of this drug. Loss of appetite and weight loss is also seen. The most common side effects as listed by Lilly are upset stomach, decreased appetite, nausea and vomiting, dizziness, tiredness, and mood swings. These are not unlike that associated with the stimulant medications.

Lilly states in its press release in regards to Strattera's introduction: "It's not known precisely how Strattera reduces ADHD symptoms. Scientists believe it works by blocking or slowing reabsorption of norepinephrine, a brain chemical considered important in regulating attention, impulsivity and activity levels. This keeps more norepinephrine at work in the tiny spaces between neurons in the brain." If we examine this statement carefully, we see it states 'it is not precisely known', therefore once again a drug is being prescribed whose effects are not fully known for a 'disorder' whose psychopathology is not yet delineated. Clinical trials for Strattera have been limited and any information on long term effects has only been studied by Lilly itself. It is interesting to note that before Strattera was actually placed on the market and had just received FDA approval that the stocks for Eli Lilly rose 6% at the announcement (CBS Marketwatch, November 27, 2002). Lilly is aware that it will profit highly by being able to market a drug as a non-stimulant (though its ill effects are similar), that is not a Schedule II drug thus less subject to scrutiny and regulation. Hemant K. Shah, an independent analyst quoted in an AP Health News Report (August 15, 2002) states that Strattera's market potential is large at a time when Eli Lilly is seeking to offset recent setbacks. , "Parents who have refused stimulant dangers because of their knowledge of the hazards involved will now be coerced to utilize Strattera being led to believe it is somehow safer because it does not fall into the category of a stimulant/ Schedule II drug.

CHAPTER 5 JOURNEY THROUGH 'MADNESS'

What is termed 'madness' or 'mental illness' is often the person's only means for expression of their being lost and confused in a world which has caused them deep hurt and pain. Such is not disease but behavior with metaphorical meaning. These individuals have received through life mixed messages and been placed into situations where regardless of the option they choose they feel damned. They seek to break out from the reality which has only caused them distress. The development of hallucinations and delusions are all metaphors for the very real demons they have encountered in disordered society. The inner mind, the voice within us, becomes amplified, and becomes 'possessed' with the demons coming forward from the trauma and distress which has been encountered. Rebellion against the system of things becomes self-destructive as the person seeks to send a message to the world of their distress, but it remains unheard. Each coping mechanism that has been employed has often led to failure and not brought them out of the unlivable situation that is their life. However, the catharsis of this pain and grief can go in two directions- it can be misery and existential death, or it can be transformative. Through the pain and struggle, through the breaking out of the 'typical reality' one can journey through various modes of altered consciousness. Many deemed 'mad' speak of the supernatural. They have sought every attempt to reach out and create meaning. If they can be helped by a loving, supportive network to navigate through this state of confusion and the various realms of altered consciousness towards rebuilding and reconstructing a life of meaning, then they can come forward to a recovery that gives them valuable insight about human nature and who they really are and the reality of the impermanence of this life and the world around us. They will find that suffering is an inevitable, and it that suffering is the state of the world which is mired in greed and attachment. The one deemed 'mad' for once has accomplished a rare task- they have completely detached. But this detachment is only from the typical standards of the world. They remain haunted by the visions of their previous life. They cannot escape it, and thus they become anxious and paranoid that something or someone will pull them back to that painful existence. At times, rage comes forward as the reaction to challenges, but who would not be outraged if their voice was suppressed and they became the scapegoat for the problems of their families or those around them? Those deemed 'mad', feeling always alone, depart to a world where they remain alone from people, yet may create for themselves beings who give them comfort and solace. This is really the end of their search, to simply be accepted and loved. But here too lies a problem, for when their lives have been devoid of love and they receive unconditional love, it becomes like an overwhelming fire that consumes them. They have never been loved, so how can they respond to an outpouring of love? When all they knew was that oppression and coercion was said to be because 'we love you', when 'love' really was only about control, how can the person then understand genuine love? Once again, the confusion sets in. To reach the person who has been deemed 'mad', we cannot overwhelm. Our sincerity will not be enough, for there trust has been shattered time and time again. It is only through entering their world for what it is, by joining in, and learning to speak the language, can we ourselves begin to understand the experience of these individuals. It is only by this joining in that the person may have the chance for their journey known as 'madness' to reach a transformative ending towards recovery.

CHAPTER 6 HEARING OUR SERIOUSLY DISTRESSED ADOLESCENTS

The distressed adolescent often has feelings of abandonment, emotional detachment, withdrawal, and isolation. These children begin to develop an intense anger directed towards an adult society that they feel has hurt them and does not understand them. Parents need to learn how to build relationships with these children and this can be accomplished through a process of emotional coaching, of allowing the child to express their feelings without judgment while providing clear guidance, limits, and expectations. It is often inconsistency and lack of clear guidance from parents that further the struggles for these children who then begin to seek guidance from misinformed peers.

These children need love, affection, and a non-judgmental atmosphere. If love does not come from a meaningful and sustainable adult relationship then it will take on a new and contorted character where the concept of 'love' comes from trying to be accepted by peers (even if they be negative ones) as the child will know that they will find a source of non-judgment and will be 'liked' even if it causes their eventual self-destruction. Affection that is not provided by adults, who should be responsible, is then replaced by irresponsible sexual activity where the teen not only seeks for pleasure in a world that often provides only hurt, but feels once again that through sex, they can find a sense of acceptance and supposed emotional connection.

Some teens are so hurt and are suffering from the adults in their lives and the chaotic environment in which they dwell, that they turn to 'radical rebellion'. This can be seen with those children who are 'cutters' as well as those into such movements as punk and Goth. With cutters, the emotional pain and trauma they have experienced is so intense, that their mental anguish manifests itself physically through the act of cutting. For the Goth teen, who dwells in a world of emotional darkness and frequent experiences of despair, once again, this mental anguish displays itself in physical signs through the wearing of dark clothing, dark objects, body piercing and fascination with things associated with death. There are also those teens who involve themselves in gangs as they are seeking a sense of connection with a 'family', even if this 'family' causes them to engage in dangerous behavior. The desire for a connection with someone who they feel will accept them outweighs their thoughts any sense of danger or risk.

Teens are seeking autonomy, but they must be taught by responsible adults that this autonomy they desire also comes with responsibility. Many teens who are distressed feel that they are controlled and are criticized. Rarely, are distressed teens positives and strengths accentuated but teachers, parents, and others frequently focus on the negative. The child enters despair and has no motivation or drive to change because they have been taught by the adults around them the attitude of 'why bother' and the feeling that they are without worth.

Parents and others must stop looking at the child as the 'problem' or try through various means to uncover some 'hidden problem' or try to blame the problem on others. If the parent can be honest and introspective, no matter how difficult and even painful that may be, they will find that there are ways that they can help alleviate the suffering of their child and they may even uncover that there were ways they contributed to this suffering. This does not mean the parent must wallow in guilt, but rather to recognize the things that must change for the teen and the family to have a more harmonious relationship.

CHAPTER 7: RESOLUTION FOR OUR CHILDREN IN CONFLICT

In counseling children and adolescents, there are key traits that the therapist must possess for the therapy to truly be effective. First, we need to avoid adultism. Adultism is a very negative method of addressing our children's needs. The therapist's orientation cannot be to have the answers and solutions or to in some way dictate a path for the child. The child needs the opportunity to feel unconditional regard, respect and dignity and to be a co-participant in the process of emotional healing. Children and adolescents need to be seen as individuals who are responsible, creative, unified social beings whose behavior is purposeful and goal directed (Prout and Brown, pg. 111). It is necessary to have 'social interest" in our children. I am in agreement with Adler who felt that children should be viewed as capable of addressing problems in a rational manner and is able to find ways to solve them. Adler de-emphasized punishment and authoritarian control in favor of relationship, mutual understanding and reason (Breggin, 2000). As behavior occurs in a social context, it therefore has social meaning, it is necessary for parents to build relationships with their children, to model and aid in their child's emotional development and well being (Prout and Brown, pg. 111). What can be defined as 'misbehavior should be viewed as our children seeking to speak to us about their emotional needs. Dreikuss (1948) examined four basic goals at the root of misbehavior. First, he felt that attention seeking was one facet of misbehavior. The child feels insignificant in some way and is yearning for the attention and support of adults. In today's society, we are faced with critical dilemmas in within the worlds of our children- divorce, single parent homes, domestic violence, and financial distress, among others. Each of these dilemmas divert our attention from what should be our cherished resource- our children. Some of the children I have worked with who have received labels of ADHD or ODD have sadly had no actively involved paternal figure. The more appropriate term in this case would be 'dad attention deficit'. Because of the lack of attention, they became children in conflict. The second area that Dreikuss examined was that of power, where children feel denigrated in some way or feel that they are lacking of a voice. Particularly with adolescents power struggles can erupt as the adolescent seeks to assert his or her autonomy and become a person of their own. These power struggles with our children and adolescents can become worse when adults use inappropriate modeling and try to subdue the behaviors and actions of the child through threats, humiliation or physical punishment (Prout and Brown, pg. 115). Third, children begin to develop the idea of revenge. This comes about by feelings of being treated unfairly and becomes an attempt to rectify a wrong in their life by lashing out at another. Often, this revenge is directed towards adults or towards the society that they represent. This is one factor in juvenile delinquency, as these adolescents seek to rebel against a society that they believe has hurt them and oppressed them (Prout and Brown, pg. 116). Lastly, Dreikuss mentions the state of assumed disability. Here the child no longer presents with serious behavioral issues but begins to 'shut down' they become less motivated, they look upon life as a series of defeats and failures and they feel that they just cannot make their way out of the quagmire. Today's mental health profession has been commandeered by biological psychiatry. Biological psychiatry has sought to look upon the behaviors and emotional world of our children and adolescents as the result of 'broken brains' and views children expressing distress as somehow disordered.

Stigmatization is another factor to think about in aiding children. This form of therapy seeks to limit the effects of stigmatization. If a child is assessed with a particular label and become aware that they are somehow 'apart from the norm' and perform at a lower level behaviorally, academically, emotionally, and/or socially, this creates a definite problem in issues of self-esteem and confidence. The therapist must not encourage the stigmatization and this is what will occur if the therapist does not keep his role as trainer and allow the child the ability to move into the reflective stage. The child will see that he is 'apart from the norm', will develop dependence on the therapist, and without a trainer, will never learn or utilize the schemas and strategies to enhance his performance. The child will continue to function only on the routine level, and there is the possibility that as he sees himself continuously being labeled as 'apart from the norm' that he will reject certain social boundaries, thus regressing to the random stage, where behaviors become spontaneous and sometimes unpredictable. It is important to emphasize the strengths and creativity of the child and to capitalize on these things as a way to boost additional confidence and give an 'extra shove' to the reflective stage.

How do we define successful treatment with our children? Is it our ideal or the child's ideal? Our goal in therapy should only be to aid the child in avoiding that which is harmful, of realizing their creative strengths, and being able to find meaning in an ever complex and complicated world. The mental health profession needs to return to the principles of compassion and empathy. Positive regard is one of the most critical needs for our children (J.O. Prochaska, 1979) It is through being treated with positive regard that guides how our children will view themselves. When children do not receive unconditional love, it is then when distress occurs, and what is seen as misbehavior is only a manifestation of this distress. Children need to have a voice, to feel that they can speak to us as adults and know that they will receive validation. We need to share in the experience of our children, to experience their feelings and to be able to reflectively listen to what they share with us (Bohart and Todd, 1988, pg. 132). Baker (1996) has stated, "Children benefit from knowing that someone cares and is trying to understand that someone cares and is trying to understand their circumstances. They respond best to counselors who provide support and understanding by creating facilitative mutual relationships. In sessions with children one of the first steps that I utilize is dramatized grievances (Breeding, 2000) and the use of mutual benefit agreements. This is similar to the fourth component in Glasser's reality therapy, which seeks to plan responsible behavior (Prout and Brown, pg. 312) I will first interact with the child in an one on one setting and discuss their interests, strengths, and allow them to vent about areas of distress. From here, we will discuss what ideally they would like to be different, what should be expected of them, and what their parents could do to support them in their goals and in achieving that ideal. The parents are then consulted in much the same way. By the end of this first session, an agreement is drawn between the parents and the child that reflects their positive shared energy and attempts to create a plan that will foster and improve their relationship, alleviate distress and stressors, and encourages that each have a voice and are able to reflectively listen and collaborate with each other. It is my view that the therapist's role is to be that of a trainer by motivating, explaining, demonstrating, redirecting, and supervising a skill practice (Goldstein, 1990). The therapist functions as a facilitator in building confidence and a level of independence where the child will be able to more fully develop needed skills and utilize them for progression. Therapy thus incorporates all facets of the child's world, thus it takes into consideration not only the home environment but also the school environment and the various social outlets in which the child and family are involved. By this, the therapeutic interventions are not limited but can provide a fullness of treatment, setting goals in all arenas. Numerous dilemmas in families come from environmental stressors and it is of utmost importance to develop the 'hardiness' (Kobasa, 1979) that will allow a child to understand and cope with high levels of stress. The buffers to stress are: social support systems (Gottlieb, 1981); a sense of control over the events (Badura, 1989), and cognitive problem solving skills (Shere and Spivak, 1982). In therapy, we are dealing with the ways we process information, react to environmental stressors and situations. For in limiting environmental stressors and distractions that affect interactions and performance, one will develop the further confidence, independence, and the ability to problem solve and work through difficult issues, which may cause stress. It is necessary to understand what is dysfunctional mental processing and to be able to curb it, as it is often dysfunctional thinking in itself that causes the burden of stress and distraction (Ellis, 1993). Among the several types of dysfunctional thinking that create stress and distraction are: all or nothing thinking, perfectionist thinking, overgeneralizations, catastrophizing, and self-punishing thinking. In therapy, the child is aided in cognitive restructuring (Goldfried and Goldfried, 1980), to first realize what cognitions cause an emotional arousal, to examine irrationality of their particular dysfunctional thinking and then to begin exercises to help the child develop healthier cognition.

One can aid the exploration and experiences of the child, helping to develop new schemas and restructure thought to be able to find solutions through strategizing. In therapy various exercises can be employed to aid in this strategizing. These exercises include ways to apply common rules and themes to situations, apply decision trees (where the child maps out a problems, and evaluates various choices and their ramifications), use metaphors and analogies, and to be able to brainstorm and set realistic and attainable goals. Feeling ill equipped and not able to strategize, the child will sometimes develop frustration to the point of completely shutting down. The inability to strategize has nothing to do with intelligence or the lack there of. Rather, the child has not fully developed the ability to be able to engage in effective problem solving. I see the behaviors of children in 3 'stages'- random, routine, and reflective. They exist in a hierarchy with reflective being the highest stage of cognition, however with the necessity of all three existing for a child to have healthy mental processing. The random stage accounts for behaviors, which require no thought and are often produced by impulse. The routine stage exists mainly of socially learned behaviors which the child has become trained to know and utilize and which soon requires little thought. The reflective stage is where the child begins to take on a true quality of independence, to be able to think for one's self, to strategize, and develop new schemas. This is a stage of exploration and creativity, which boosts confidence. Children who are impulsive and have difficulty in social settings remain mainly locked into the random stage. Those children who may have few problems behaviorally in various social spheres but who either have poor decision making skills, and/or are unable to remain focused on higher tasks can be said to be caught in the routine stage. They have developed various social skills from others but have yet to be able to fully mature into their own person, they retain dependence because they are yet to be able to develop a cognitive map for themselves. Parents and teachers can often without even being aware contribute to the child remaining in the routine stage by expecting mere conformity, talking down, and/or coddling the child. As stated prior, the therapist must be a trainer. Thus, he or she must only facilitate the movement to the reflective stage and assist in the development of the child's skills. The therapist must often remain on the sidelines 'cheering' the child on and he tries out his new concepts and begins to work through strategy. The therapist (and parents as well) may fear- "What if the child fails? Won't this lower his confidence?" First, just as a football team with a good coach will perform better, it is unlikely that a child with a therapist as a good trainer will fail. Second, it may take some time for the child to be able to integrate the skills he has learned, but it will show to be a better confidence builder to have the child accomplishing independently rather than with the continued support and dependence on others. The child must be allowed to move from the routine to the reflective and this is a difficult process. The therapist must only facilitate this movement. When the child's performance is lower than expected and goals are not being met in time, this is a time to regroup. The failure of the child is more the failure of the therapist to be an effective trainer. This is where the therapist can learn much as well. He or she can explore with the child- what strategies were used? What level of frustration was present?- and thus plan out a course of action and a restructuring where the child can learn from mistakes. In my therapeutic session, I have also sought for them to be intensive and require the child to also complete certain work and exercises in between sessions. A detailed journal is kept in which the parents, teachers, and others involved in the child's life take active participation recording areas of progress and areas where improvement is needed. The journal also allows the child the ability to conduct self-assessment and express his or her emotional needs. The child can examine what are activating events for him that lead to certain beliefs which may cause certain consequences and reactions, among them distress (Prout and Brown, pg. 252) Therapy is collaborative, and works to build bridges and foster relationships and dialogue between all those involved in the child's life. This journal becomes a framework for designing therapeutic sessions that meet the particular and individual needs of the child and family. In helping the child to gain better insight into his or her emotional world, I have borrowed from rational emotive therapy the use of a "SUD" (subjective unit of distress) (Wolpe, 1990) can be employed in this therapy, this is a means by which the level of frustration, and the level in which the child finds interest and/or 'fun' in certain areas can be measured. An example would be to have the child construct a 'thermometer' and before he begins a particular task to measure what he feels his level of frustration is. One can begin to work with the child in strategizing and working through the task as a trainer and partner, and then once again have the child use this 'thermometer' to rate his or her level of frustration. This will often show the dysfunctional thinking and help the child to realize that he or she is able to complete the task, thus giving encouragement and building of confidence. Children's attention spans are different, so it is important to design the interventions in a way where they do not become overly grueling, and where the goals can be effectively met in a way that will keep the child interested and focused. Thus, interventions need to be simple and straightforward and geared to the level at which the child is presently at. A part of this therapy is also to get the parents and/or teachers to realize what dysfunctional thinking they may hold to that can hinder the child. Particularly, parents may hold to expectations for the child that are unrealistic. Teachers may also expect too much from the children and also wrongfully believe that a child's performance means he or she does not have the potential or aptitude to complete certain tasks. A teacher may in some cases without even realizing it develop an attitude towards the child where the child is 'looked down upon' or treated differently from other children. The peers may mirror the teacher's attitudes and peer interactions become affected. Within therapy as well, the child learns anger management and essential skills in conflict resolution, how to be able to clearly define a problem and express it in terms of needs and 'doables', he or she begins to learn how to develop 3 part assertion messages to be able to clearly and effectively express feelings. The child learns reflective listening, which can play a key role in relationship building as well as giving children with attention problems that ability to learn how to listen more carefully and be able to follow and repeat back instructions. The Greek philosopher Epicletus stated, "People are not disturbed by events themselves, but by the views they take of them." As Stanley Greenspan, MD states, " The goal of mental health treatment is to help the individual move toward development appropriate to his or her age...I have seen that the mind grows from certain type of experiences. Insights, advice, and behavioral modification strategies do not provide these experiences." Shared experience and relationship are key elements for healing. The therapist can then begin to help the parents and the child to develop mastery over emotions and begin a process of growth appropriate to each's capability to assess and restructure emotions and relationships. The physician and anthropologist Melvin Konner states, "You can get short term gains in self-esteem and continue to lose ground; or you can try this theory: that self-esteem can also come from making a great effort, from facing uncertainty and overcoming obstacles that we were not sure we could meet, from doing our level best. You may have to struggle with a child to get her to overcome her doubts about herself, to dig in and really try. It's a risk. But only by taking it do you get to see that smile- no, it's a grin, really, and the face is open in a hint of astonishment- that breaks over the child's face as she slams the pencil down on the page and says in a thrilled, surprised voice, "I did it!" As Greenspan comments, each child exists at different levels, but when items are broken down in a way that is understandable to the child's capabilities and strategies are given by the therapist and others to work through problem solving, mastery can be accomplished. This is why it is necessary for the therapist to be only a trainer and to step back from being actively involved in certain tasks the child is to carry out, the therapist is only a coach, an observer to help the child be able to develop mastery as an independent person and to be able as Konner states, to build self-esteem in knowing that their accomplishments come solely from their own efforts. Parents have an immense responsibility and must be partners in the therapeutic process for progress to occur. Parents cannot be blamed or denigrated. As Breggin (2000) states, "I avoid blaming parents, and in particular I help them overcome their feelings of guilt. Guilt is a painful, debilitating emotion that rarely motivates us to be more loving and concerned. It can be useful to point out to parents how they may be contributing to their child's problem, but it's always more useful to provide new directions in how to discipline, care for, and educate the child. Often the problems don't especially lie in the family. Nonetheless, the parents remain most responsible for discovering and correcting the source of the problem. In teaching principles of discipline to families I avoid the typical behavioral modification approaches. In order to train dogs, we often will provide a treat to the dog in order to get it to comply with a command such as 'sit', 'shake hands', etc. However, in this the dog often becomes reliant on the treat to produce the expected behavior. Those who wish to use behavioral modification techniques are doing the same with our children. The child is forced into conformity through the use of rewards and praise but never learns independently how to make decisions for himself and conduct self-assessment. Mary Budd Rowe, a researcher from the University of Florida determined that children that were praised lavishly began to be more tentative and unsure of their responses, waiting first for adult approval before making a response. Behavioral modification takes way the free thought process of children and also affects their level of esteem and causes them to become more dependent. Even praise is a judgment and children to not desire to be judged, it is more effective to acknowledge that a child has completed a task and allow him or her to reflect themselves on how this made them feel rather than to impose our own feelings on the matter. Praise is conditional, it implies to the child that we will take away attention if there is something that does not meet with our immediate praise, and thus they loose interest and become less motivated (Kohn 2000). Children need our attention even when their activities and actions may not meet with our approval. Our response should rather be the following as Alfie Kohn points out- say nothing- allow the child to determine for themselves the result of an action; say what you saw- do not make judgments but merely acknowledge the action itself; Talk less, ask more- determine what the child felt rather than imposing what you felt. Discipline should only be used with children in the cases of self-defense, where the child is infringing on the rights of another by harming you or another. Otherwise, discipline will create rebellion and dissension and prove completely ineffective. Those who support behavioral modification techniques often utilize behavioral charts designed mainly by the parents where the child is forced to compliance once again by the rewards system and its problems as I mention above. According to a study by University of Rochester psychologists Edward Deci and Richard Ryan, rewards simply control through seduction rather than force. What I have seen in my experience to prove far more effective and reduces the need for parental 'nagging' is to have the child involved in determining what he or she feels should be expected of him or her and then to see how much is accomplished but not to 'nag'. I have seen children more interested in being helpful and accomplishing various tasks when it becomes their idea and concept. Time outs have often been used for discipline and redirection, but do not solely get to the root of the problem. Time-out needs to be used at the first hint of misbehavior and not after multiple prompts and cajoling. There should be set for each week target behaviors that are to be dealt with and which the child is expected to collaboratively work upon. After a time-out, the child should be required to give a reason why they were there and a plan for corrective action. Unless we require our children to be reflective, they will not learn self-control. Often the label of ADHD becomes a means of parents being able to exonerate themselves of taking responsibility and for children to never learn self-control. It is important also to have daily 'time in' periods to explore why he or she is feeling the way he is, why he feels the need to act out, explore completely from the child's point of view and rely on the child to determine what may be solutions to issues where the child might be infringing on the rights of others. The use of the 'time in' occurs after the child has de-escalated. In addition, quality time and interaction needs to occur more frequently with children. So often adults become caught up in a fast paced world, and their children become lost and ignored. One should respect the space of the child, allowing them to vent and make use of dramatized grievances (Breeding, 2000) if necessary. Intervention at this point should only occur if harm is to occur to the child or another. Once there is de-escalation, the therapist and/or parent can then begin to process the incident with the child. We should also remember that we should never disempower the child nor use our own power to force conformity on a child and make him do what others are doing when he or she may not be in the proper emotional state to be compliant. We should evaluate as well just what it is we are expecting the child to be compliant to and whether this is reasonable or not. If a child is experiencing some distress and is more withdrawn or uninterested in other activities that other children are doing, we should explore with the child feelings but not force the child into activities that he may not be in the state to perform in. If a child is experiencing an emotional distress, we should not be thinking immediately of discipline or redirection, but rather how can we support this child, how can we explore what this child is feeling, how can we be empathetic? I have introduced to families the concept of 'selective attention'. Often I have seen that whereas children certainly need attention and validation that validation is often given to the wrong things. For example, a child becomes angry and upset because of a consequence given by the parent, he may yell out, "You are mean, you are not my mother" among an array of other comments. The parents often respond to these comments and validate it with some sort of feedback such as, "Ok, then I am not your mother" or may come across with various threats. This only causes a cycle of further comments being spewed back and forth between the parent and child. The parents for that moment need to control themselves and be able to not respond, to allow the child to have his moment of expressing his or her anger or frustration at the moment, and to be able to deescalate. To provide any type of feedback, positive or negative during this time, is not going to aid the child in being able to process. Once the incident is all over, the parent can then have a one on one time with the child to discuss the incident and to work on better coping strategies so that the child can learn to manage frustration in a more effective means. Children cannot be expected to always act mature and responsible, when they are frustrated and angry, it is only obvious that this will come out in some way. The child cannot be blamed for this, and the adults need to take responsibility and assess their roles in how a child responds to frustration, many of the their responses are patterned on how they see others, particularly adults reacting. Aggression will breed aggression. Even when a parent engages in what may seem minor verbal aggression, this itself has an effect. Parents need help in being able to cope and manage their own feelings so that they can effectively model appropriate behavior for their children. Sadly, in today's society, feelings and behaviors are suppressed by medications, and as a result children and their parents as well never truly learn the skills to manage frustration. The label of oppositional defiant disorder has been used for children who express their feelings in ways that may not always seem acceptable, but the fact remains that based on the criteria for the disorder, every child would receive this label at some point in their lives, as children develop and cope in different ways, some more effectively than others. The goal is then to be able to work on these coping mechanisms, not to suppress feelings, which leaves them buried only to resurface again at some future point. This is where the model of therapeutic intervention I have outlined helps to truly get to the root of the problem. Many of the families I work with in a therapeutic setting are disadvantaged, and as a result there comes many environmental factors and stigmatization that has a direct correlation many times to behavioral issues. Few want to really look at the need for environmental change as being crucial and are satisfied with placing labels, which blame the child and stay with the child throughout his or her development. I am convinced that psychotherapy cannot be effective without the therapist being willing to also help the family find the support system and resources that may lie outside his particular therapeutic role. If the therapist is to be looked upon as genuine, sincere, and empathetic, then he must often take on additional roles as an advocate, a trainer, and a trusted individual. Some will say that by doing this we are 'over investing' ourselves and becoming too extensively tied and caught up in things that will cause the therapeutic process to be clouded. I see the opposite to be true. We can never invest enough into the lives of children. We cannot expect to design a treatment plan and have sessions with a family where there exists a chaotic environment. Many families see therapists as a part of a system of folks who come and tell them what they need to do, get paid, and leave. Many of these therapists focus solely on a 'book' or what the 'experts' have to say. The therapist cannot present such an attitude towards families or it will be viewed that they often ignore what the family and child has to say, discounting the fact that they are part of the team, and more than this, are the most important part of the team. The family can only be moved to healing when they feel validated, loved, supported, and can build a trust with the therapist that they can share things at the deepest levels. Let me give an example- a child I work with had some difficulties with self-esteem and we were seeking to boost confidence. This child took great pride in the fact that he would be participating in a school performance and invited everyone he knew including the therapeutic staff. I went to this performance, and his face gleamed after the performance and more so when he saw his family and friends. When he and his family saw me there, this showed to them that I am not in the field merely because it is a 'job', it showed them that when I stated in the treatment plan I was going to boost this child's level of confidence, that I meant it. It worked, the child has progressed and we have capitalized on his own strengths to encourage his own process of healing. Was this an over investment? Possibly so, but did it help further progress? Yes. To me, that's all the matters. I believe that therapeutic intervention must be short term, it troubles me to see how that some children spend ages with therapists and the situations drag on and on with little if no result. If it means I must invest additional time (for which I may not be paid), it is well worth the 'over investment' to see that the goals of treatment are really being met and that we are not wasting time along a dead end road. Having a therapist or others involved with a child in itself can carry a stigma. If a child has someone in the school working with him, though this certainly may carry some benefits, the fact that the other children do not have such a person and see the child with one may cause them to view the child differently. The teachers themselves may treat the child differently as a result as well, and may without even realizing it, encourage stigmatization and the breakdown of peer interaction. Not all children respond in the same ways, and the teacher and others working with the child must be careful not to talk down to the child but must respect his uniqueness, his sovereignty, and his dignity. Often I have been told that my role is solely in the home, and to focus too much on other spheres would once again be an 'over investment'. But, is a child's life really that compartmentalized? The problem with some therapists today does not lie in 'over investment' but 'under investment'. A systemic approach is necessary and the utilization of 'do whatever it takes' philosophy as well as connecting the family to community resources and the realization of their own strengths (Instead, they look upon their role merely as a simple profession, develop the 'organized psychology' mentality and convey without maybe even knowing it a message to the child and families that they really lack compassion and empathy. When we examine most emotional disturbances we will find that largely the problem is the inability to have a sustaining relationship or a relationship that has been broken. Individuals then respond in various ways and may adopt behaviors which either push people away from them or respond in a way that shows that they have an intense fear of not being able to trust and forge a new relationship. This is why love and affection are at the key of any healing. A psychotherapist has a great burden to earn the trust of the child, which has been shattered by circumstances. Children whose worlds have been 'shattered and whose trust violated may experience many conflicts, among them depression. Depression is an emotional state, not a brain disorder. It is a state where individuals are unable to develop a coping strategy in regards to life. At some point people have been depressed, it is a natural part of our existence to have times where we are overwhelmed by life's problems. A problem arises when feelings of hopelessness and despair reach a point where the person is not able to effectively deal with them and remain in a darkened emotional state for a significant period of time. "It is a mistake to view depressed feelings or even severely depressed feelings as a 'disease'. Depression, remember is an emotional response to life. It is a feeling of unhappiness that involves self-blame and guilt, a sense of not deserving happiness, and a loss of interest in life (Breggin, 2001). Many who exhibit feelings of depression are being encouraged to take 'anti-depressants'. The fact remains that these medications do more harm than good. "The term 'antidepressant" should always be thought of with quotation marks around it because there is little or no reason to believe that these drugs target depression or depressed feelings. In fact, (there is) considerable evidence that these drugs have little or no therapeutic effect on feelings of depression (Breggin, 2001). "There is no magic bullet medication that can target depression and cure it for you (Breggin, 2001). "A caring therapist, a loved one, or a devoted community such as an extended family or church can be lifesaving. But when a doctor spends fifteen minutes with his patient and prescribes a drug, the sense of aloneness and isolation is likely to be reinforced. The very idea of turning to pills instead of people can add to the feelings of despair and hopelessness. In short, it is depressing to believe that an improved life is best achieved by taking a pill (Breggin, 2001). Those who are depressed need the support of individuals who will support them and accept them for who and what they are, encouraging their growth and gradually taking them through the process of adaptive responses so that they can develop the coping mechanisms needed to deal with life. These individuals need a reason to keep living, they need to be able to set for themselves attainable and realistic goals, otherwise they will continue to be shattered by disappointment. Many times just a simple listening ear can work wonders in promoting the emotional health of depressed individuals. These people need a daily routine that does not trap them into continued feelings of depression. They need a time to be reflective. Often a spiritual support system can be very effective, as depressed individuals need the support of a community and a map that can trace out answers to life's overwhelming difficulties. These individuals need to be able to examine and confront past events that led them to the point of depression and hopelessness. They need the aid to rethink their thought processes and develop new strategies. Breeding (2000) has stated, therapists, parents, and all adults must begin to see children through the eyes of delight. It is an important intervention for us to begin as well to realize that in we should all work for social justice. This is where much of the emotional distress of children originates from. We should strive to improve education, community resources, and to once again hear the voice of our children and respond.

CHAPTER 8: AUTISM AND ENTERING THEIR IMAGINATIVE WORLD

In dealing with children with autistic children it is all about relationship. These children are within a realm where they feel and respond much differently than others. There has been much focus on trying to eliminate certain behaviors or to evoke particular responses in children which actually become rote and repetitive for them without context. One of the goals in aiding these children should be in helping them find meaning. In order to do this we must be willing to not look at the child as broken, unable to respond, or even unable to communicate. These children DO communicate, however they are not always able to manipulate their senses to communicate in the typical ways of other children. As a result, they can become easily frustrated and trapped. The therapist must enter their imaginative world and learn to communicate in their language.

Stanley Greenspan gives an example of a child who initially went to a psychologist who engaged the child in repetitively placing pegs in a board or trying to find beads hidden under various cups. This was supposed to be a measure of the child's intelligence and abilities but it proved ineffective. The child constantly hurled the pegs to the floor. A different psychologist took a unique approach in having the mother participate with the child in a series of interactions. First, the child began grabbing the nose of the mother. Rather than redirecting the child and seeking to have her refrain from the grabbing, the mother responded with a 'toot toot' noise and then allowed her to do it again responding with a new noise. The mother then gently touched the nose of the child and the child to the amazement of the mother smiled and let out a noise, "mo mo". The child had indeed communicated but in her own language. The mother and child had made a real connection. This showed to the psychologist that this child's cognitive development was within a normal range and here was a child who wanted to exert some control over her surroundings. Over time, the communication increased, and the mother was able to have 'pleasurable' discussions with her child that prior had never existed (Greenspan, The Growth of the Mind, 1997, pg. 8-9)

    Children with language difficulties need to have emotional and social supports. Unless these are more fully developed, the language will be fragmented and lack meaning (Greenspan, pg. 32). Before language development can come, improving the understanding of non-literal and non-verbal communications need to be worked upon. There are 6 main milestones for children: self regulation and interest in their surrounding world; intimacy; two way communication; complex communication; emotional ideas; and emotional thinking. In Greenspan's floor time model the first goal is to encourage attention and intimacy which helps in the further development of the first two milestones. The parent will actively participate in a period of play therapy engaging their child in creative play allowing the child some direction over the course of the session and taking interest in their activities as well as providing encouraging feedback. Self-regulation becomes difficult for some children because sensory stimulation can be so overwhelming or their attention may wander (Greenspan, Essential Partnership, pg. 8). Difficulties in intimacy occur because the child is not able to effectively read the cues being given. Often times the children will have an easier time with adult relationships because adults are more able to adjust their cues to the level of understanding of the child whereas this does not always occur with peers. A part of reaching out to these children and guiding them in the intimacy milestone is to provide them opportunities to interact with peers and to have them be able to relate back what the other person is stating and feeling. Making use of social stories and role plays can be helpful in aiding the child in understanding the feelings of others as well as their own feelings. A social story is a device used where a make believe dialogue is constructed and the child is asked to fill in the gaps. "A social story is a story written to specific guidelines to describe a situation in terms of relevant cues and common responses (Gray & Granard, 1993). The use of comic strip conversations can also be employed. "A comic strip conversation is the genuine 'art of conversation'. This approach incorporates the use of simple drawings and color to illustrate an ongoing communication. This provides additional support to (children) who struggle to understand the quick exchange of information in a conversation (Gray, 1994). An advanced form of the social story is what is termed the 'thinking story'. "Thinking stories demonstrate the variety of possibilities as to what people may be thinking when they make certain statements, or when they display certain behaviors…Thinking stories follow a specific, structured format, using picture symbols from Comic Strip Conversations to define and illustrate the abstract concepts covered in the story (Baron-Cohen, 1990, Dawson &Fernald, 1987). The person or therapist using the social story can help guide the child through and the use of feelings charts can also be a beneficial aid. To reach the milestones of two way communication and complex communication, it is important within the sessions that the parents have that they utilize a dialogue with the child, help guide them to use their face, emotions, hands, to convey their needs and desires. Encouraging the child's imagination and creativity will help in the development of the complex communication as they begin to move towards problem solving. Lastly, it is important to work on logical thought, being able to take the things they have learned from the parent's coaching and to actually be able to convey some insight and understanding of the world.

In the play therapy sessions, it is important for the parent and/or therapist to actively participate. The purpose should not be to entertain the child, but to interact with the child. Seek to draw near to the child, but this should not be forced, allow the child to express themselves at their particular pace. Use lots of gesturing and cueing and become a part of their imaginative play, allow them to show and teach you something about their world. It is important to not just tolerate their feelings and certainly not be dismissive of them, but allow the child to express their feelings openly being able to distinguish feelings from behavior. Don't be afraid to challenge the child in new skills, they will be eager to learn as long as the challenge is not forced. From time to time, you will notice that these children will become obsessed with routines or repetition, so in the play do something to break the routine or repetition. If a child is repeating a certain topic or action, do something entirely different that will focus their attention elsewhere. Do not be repetitive in your directives and follow a plan of rote learning, allow the child to explore and display what they do know. It is important to ask open ended questions, let the children explain to you. Find out what these children find meaning in, and seek to have them tell you why. Don't judge or evaluate their answers, but be a listener. Help the child to brainstorm new ideas, and particularly when conflict arises, let them be able to perform some self assessment, sit as a partner as they develop adaptive responses and utilize them. Don't be afraid to allow the child to fail from time to time, they will learn and gain insight from their trial and error. When the child is expressing certain thoughts and feelings, help them to be able to label what it is they are expressing (Greenspan, Essential Partnership, pg. 20)

There are key social behaviors as they relate to relationship building that should be addressed with the high functioning autistic child. The first is entry skills. This refers to how the child joins a group of children and whether or not they seek to include other children into their play. The therapist can help serve as a coach for entry skills and encourage scenarios where the child will have opportunities to exercise the skill (Atwood, 1999) Next is assistance, whether the child recognizes when to seek help from others or to provide help to others. Social stories can certainly be utilized in this situation. An example of a social story as given by Dr. Tony Attwood (1999) that applies to this skill is as follows: Sometimes children help me. They do this to be friendly. Yesterday, I missed three math problems. Amy put her arm around me and said, "Okay, Mary" She was trying to help me feel better. On my first day of school, Billy showed me my desk. That was helpful. Children have helped me in other ways. Here is my list: I will try to say, Thank you! when children help me. Another example of a social story is: My name is Josephine. Sometimes, children help me. Being helpful is a friendly thing to do. Many children like to be helped. I can learn to help other children. Sometimes, children will ask for help. Someone may ask, 'Do you what day it is today?' or 'Which page are we on?' or maybe something else. Answering that question is helpful. If I know the answer, I can answer their question. If I do not know the answer, I may try to help that child find the answer. Sometimes, a child will move and look all around, either under their desk, in their desk, around their desk. They may be looking for something. I may help. I may say, "Can I help you find something?" There are other ways I can help. This is my list of ways

I can help other children: Children like to be helpful (Atwood, 1999). For younger children the use of the Mr. Men stories (such as Mr. Nosy, Mr. Grumpy) by Roger Hargreaves can prove useful.

The other skills which need development include receiving and accepting compliments, accepting and receiving criticism, accepting suggestions, reciprocity and sharing, conflict resolution, monitoring and listening, empathy, and learning to ending meaning how to provide closure to an interaction. For conflict resolution skills I recommend the use of Weeks's 8 fold model. In this model one first provides and effective atmosphere for the discussion and resolution of the conflict, clarify perceptions, focus on needs, build shared positive power, look to the future and learn from the past, generate options, develop doables, and make mutual benefit agreements (Weeks, 1992).

Within the education system is a great misunderstanding of high functioning autism.. These children cannot be placed in an autism classroom as they are too high functioning. These children can be challenging and some teachers and school administrators are afraid of taking the necessary steps to insure these children's success. Partial hospitalization becomes an easy out for the school districts. Teachers need to be able to build a relationship with the child and recognize their strengths, being respectful of the child's personal space and boundaries and always speaking to the child in a calm and collected manner. "Teachers need to have a calm disposition, be predictable in their emotional reactions, flexible with their curriculum, and see the positive side of the child (Atwood, pg. 173) Some teachers see that these children will rock in their seats or move their hands or feet and look at these children as being disruptive in the class. The rocking behavior is a way that the child 'grounds' themselves, it is comforting for them, and is not a behavior to condemn the child for nor one that can or should be eradicated. If it appears to be a disruption, the teacher can provide a place for the child to be able to have a break until they feel they are more calm. School administration must understand that for the child that sensory stimuli can be very frustrating, and sometimes these children may need brief periods away from school that allow them to regain some emotional stability. Such absences should be written as allowable in the IEP and should not be treated as truancy situations. The size of the classroom is paramount for these children. "Open plan and noisy classrooms are best avoided. The children respond well to a quiet, well-ordered class with an atmosphere of encouragement rather than criticism. Parents find that with some teachers the child thrives, while with others the year was a disaster for both parties. If the teacher and child are compatible, then this will be reflected in the attitude of other children in the class. If the teacher is supportive then the other children will amplify this approach. If they are critical and would prefer the child were excluded, other children will adopt and express this attitude (Atwood, pg. 174). Once a child is in an appropriate environment with the necessary resources, this environment should be maintained. "Once parents have located a school that provides the necessary resources, then it is important to maintain consistency. Going to a new school means changing friends and the school not being aware of the child's abilities and history of successful and unsuccessful strategies." Children with Asperger's syndrome may display an unusual gait and difficulties with motor skills and coordination. They may also have difficulty with sensory stimuli so it is important for the therapist to take note of distressing stimuli and help to limit these things within their environment as much as possible. Activities designed to work on motor skills and coordination can prove beneficial but consideration should be taken as to not force a child or cause undue frustration if the child's abilities are impaired. Emotional coaching can prove effective for parents. Emotional coaching involves seeking to see the expression of emotions as a time for intimacy and teaching, providing validation to the child's emotions, and helping the child to be able to label their emotions. The parent who is an emotion coach values the child's negative emotions as opportunities for intimacy; can be patient with the child when they are sad, angry or fearful; can identify triggers; does not tell the child how to feel; does not expect to have all the answers (Gottman, 1999). There has been some discussion of a link between gastrointestinal disorders and children with autism spectrum disorders (Wakefield, 1997) Some children with autism spectrum disorders may exhibit encopresis. The child should be regularly seen by a physician if any problem arises. The child should not be punished for occasions of encopresis or be made to feel embarrassed. As pediatric neurologist Fred A. Baughman has stated, autism is a blanket term as is cerebral palsy identifying a developmental condition rather than a psychiatric issue. While those considered within the autism spectrum may display similar traits, there are diverse etiologies (Baughman, 2001). Some children with traumatic brain injury or epilepsy may display autistic traits. However, there can also be psychosocial reasons for the development of autistic traits. The term itself is very loosely used and at present the exact etiology is not fully known. I tend to look at autism as a variation in perception, yet a normal variation. These children are not defective. As individuals may be left handed or right handed, this is a variation, but does not state that a left handed individual who is in the minority is somehow defective or 'abnormal'. Rather, because children with autism have a variance in their perception, this causes them to come into conflict with the general functioning and perceptions of society as a whole. They have unique strengths but may need dome extra assistance in being able to navigate through what the rest of society typically perceives and how it interacts.

There are no medications that will 'cure' autism. Some individuals have used various medications in an attempt to control behaviors, however it must be realized that this is all that the medications are capable of doing is controlling a certain aspect of behavior by blunting certain brain functions. These medications all have serious risks. "Neuroleptics have their main impact by blunting the highest functions of the brain in the frontal lobes and the closely connected basal ganglia. They can also impair the reticular activating or energizing system of the brain. These impairments result in relative degrees of apathy, indifference, emotional blandness, conformity, and submissiveness, as well as a reduction in all verbalizations, including complaints or protests. It is no exaggeration to call this effect a chemical lobotomy…contrary to claims, neuroleptics have no specific effects on irrational ideas (delusions) or perceptions (hallucinations)." (Breggin, 1999) These medications also carry the risk of causing tardive dyskinesia or neuroleptic malignant syndrome. Tardive dyskinesia is permanent abnormal movements of the voluntary muscles. "NMS is characterized by severe abnormal movements, fever, sweating, unstable blood pressure and pulse, and impaired mental functioning. Delirium and coma can develop. NMS can be fatal…(Breggin, 1999) Common side effects of neuroleptic medications as reported by the Physicians Desk Reference are abdominal pain, abnormal walk, agitation, aggression, anxiety, chest pain, constipation, coughing, decreased activity, diarrhea, dizziness, fever, headache, inability to sleep, increased dreaming, indigestion, involuntary movements, joint pain, lack of coordination, nasal inflammation, nausea, overactivity, rapid heartbeat, rash, reduced salivation, respiratory infection, sore throat, tremor, vomiting. The SSRI antidepressants' are also a common prescribed medication. These drugs can produce akathisia, mania, worsening of depression, obsessive compulsive like behaviors, and severe anxiety and agitation (International Center for the Study of Psychiatry and Psychology Newsletter, Summer 2002, pg. 15) The use of responsible psychosocial and relationship based approaches are far better than any short term benefit that neuroleptics may provide.

CHAPTER 9: THE VALUE OF A RELATIONSHIP BASED APPROACH

In aiding children with developmental challenges, we must first realize that this requires a team effort and a strengths based approach. It is necessary to not focus on what the child cannot do but look at what the child can accomplish and build upon this. Parents can enlist the support of professionals but must realize that it is they who are the most important persons in the child's life and that furthering the development of their child is not just the work of professionals but is a collaborative effort from everyone involved with the child. It is necessary that for any interventions to truly be effective and helpful, that they must be consistent and constant. The interventions must be the same throughout all domains that the child is present in.

It is crucial for us to understand the environmental responses that children have, whether they have developmental concerns or even if they do not. If a teacher, parent, or other person has a hostile tone, a poor demeanor, a loud voice, etc. All of these things can be overwhelming to the child and can provoke a behavioral response. All behavior is purposeful and should be looked upon as so, even negative behaviors. Behaviors are a way of the child speaking to us about a distressing situation or an apparent need or desire when they may not be able to convey this to us verbally. Light, sound, and other sensory stimuli can also produce distress for a child. We need to create awareness of what in the environment may serve as triggers to distress and seek to modify the environment to make it a more comfortable and safe place for the child. We must also be cautious in how we view children. If we look at a child displaying negative behavior as a 'monster' or feel that because a child may be rambunctious at times that we must automatically resort to medicating them, then we have taken a negativistic attitude that will surely be passed on to the child. Children are keenly aware, even those with communication struggles, of adult's perceptions of them. We should look at our children through the eyes of delight and address behavioral difficulties not in terms of how we can subdue, but rather how we can meet needs and resolve conflict and remove distress.

The floor time model is of particular usefulness in working with children with communication and social struggles. For those children who are non-verbal, we can begin to introduce hand signals, moving to use of pictures, and then gradually encouraging the child to make use of words or phrases to indicate desires. It is not important initially whether the verbalizations are correct but rather that a verbal attempt was made. When a child displays such a behavior as spinning objects, in the floor time model, we would not be aversive, but rather gently introduce a new toy or object and seek to divert the child to a more productive activity. In situations of echolalia, we can say such things as 'that's TV talk', and provide means to divert this to a different means of conversing. It is important to provide the child with understandable signals and meaningful statements and phrases when we desire them to behave in a different way.

In order for children with developmental concerns to be able to integrate more into the social sphere, it is necessary that they not be isolated into situations where they are labeled and shuffled away from typical peers. Rather, they should be included as much as possible with typical peers. They may need additional support and accommodations, but how will they begin to learn important skills unless they have frequent and continuous exposure to the world around them. I have developed the use of what I term 'real life rehearsals', where we may set up a particular social scenario for a child. It may be such a thing as being able to make a purchase at the grocery store. The therapist and parents guide and coach the child ahead of time in how to go about such an activity and then have them actually demonstrate it. Social stories and comic strip conversations are very useful in conveying information as these children tend to be visual learners. Social stories can be simply made booklets that the child helps to create where a particular task or scenario is outlined with what behaviors are expected. The comic strip conversation is helpful in building empathic skills as well as reflective thinking as we ask the child to develop captions for what different individuals may state and think in various situations.

Lastly, I think it is crucial, though it may appear controversial to some, to state that children with developmental concerns can and will be benefited from a psycho-social and relationship based approach alone. Some have decided to resort to medications, and I am placing no blame or condemnation on those who have made this decision, however making a suggestion that there are alternatives and informing of these alternatives as well as the hazards of psychotropic medication usage. First, I will not argue that medications can 'work' in the sense of subduing behavior. However, strapping a child to a chair would also work in regards to subduing behavior. This would be aversive and quite possible illegal. I see little difference between such an approach and that of using psychiatric medication. The difference is that one is a physical restraint, the other a chemical restraint. When we say that something 'works', often we are not looking at the mechanism by which it works. Dr. Peter R. Breggin, MD compared the use of antipsychotic medications in children to 'chemical lobotomy' as it blunts the functions of the frontal lobes. The risk of tardive dyskinesia, a permanent disfiguring neurological impairment exists with these drugs. In addition, such drugs as Risperdal are prescribed off label and are not indicated for anyone below the age of 18 but continue to be prescribed.

It may require more diligence, effort, and patience, but I remain convinced after working with over 70 children with developmental challenges, that relationship based approaches, rather than chemical restraint, prove to be a true means to teach our children skills, to focus on their strengths, to build on their development, and to help address challenging behaviors and to address the real source of conflict and distress rather than just blunting it.

CHAPTER 10: IS THERE AN ALTERNATIVE AND WHEN COULD PSYCHIATRIC DRUGS APPROPRIATE?

I happened to share with some colleague's information about adverse events of psychotropic drugs as well as information on black box warnings. I shared the greater need for informed consent. The question posed to me was, "is there an alternative?" When I answered "yes', this was followed by, "well, we need training in this alternative." Therefore, this is where the problem lays- the level of training and how mental health professionals are indoctrinated. Only 2% of medical schools train in counseling, thus psychiatrists are solely taught to prescribe, not to counsel. The alternative lies in relationship based approaches and such programs as the Caregiver's Skills Program for children who have been considered to have disruptive behaviors. The alternative lies in educating and training parents to become more involved with their children. In addition, it requires us to take an active stand for social justice and improvements in education where the root of many children's challenges lie. It requires us to not look upon children as 'diseased', this is a fraud, but more than capable with the correct mentoring, guidance, and affection to develop into stable human beings capable of self-control. Therefore, in addition to educating on the dangers of psychiatric drugs, it is necessary for us to be actively involved in showing that indeed there is a more healthful way to meeting our children's needs.

As I have reviewed in a number of articles and can be seen in the literature, stimulant medications for children have not shown improvements in academic performance or pro-social skills in the long term and there exists adverse effects. Of more serious concern is the possibility of some children developing psychosis. Antidepressants in children have been linked with suicidal ideation in some children. Antipsychotics can lead to tardive dyskinesia.

Therefore, it is my view that with children it is very risky and dangerous business to subject them to psychiatric drugging. In addition, children have little if no voice in the matter as to how it makes them feel or if they agree to take these drugs. Therefore, psychiatric drugging of children needs to be curbed.

In situations where an actual physical abnormality can be demonstrated and where the person is exhibiting self-harm or harm to others, the use of a psychiatric drug could be considered but not as a long term solution. An example would be in a situation where a person has traumatic brain injury and is self-harming. With adults, they are able to be given informed consent. Therefore, an adult who chooses to use a psychiatric drug being made fully aware of possible adverse effects, this is entirely their prerogative. The issue is informed consent with adults.

I recall a child who came to me for an evaluation for wraparound services. This was my first encounter with the child. He entered the room staggering and swaying, barely unable to keep his eyes open. He almost fell out of his seat and his speech was slurred. When I commented to the parent that the child looked highly sedated, the parent responded, 'well, at least he's not aggressive." This led me to conclude that this parent was actually looking at this child's miserable state as progress.

All psychotropic drug effects occur by disabling the brain. Because a child becomes more subdued does not mean that there has been any real progress. The child has not learned any new skills. They have not learned self-control, they have merely been chemically strait jacketed. People do not understand the damaging effects of psychiatric drugs on the brains and development of children. They are looking at children as being more subdued in their behaviors but failing to understand how these effects come about. Children may miss a dose of a psychotropic drug and their behavior becomes dramatically worse. Then it is said that, "oh, they missed their medication, they need their medication." But think of it this way, there is little if no difference between many prescribed drugs and illicit drugs. A drug addict who does not have his fix becomes quite crazed. This does not mean the addict actually needs more of the drug, though it may appear this way and he may feel he needs more of it. So it is with children on psychotropic drugs. When they miss their dose, their behavior can become erratic. But this does not mean they need the drug or need more drugs. It means that they are having a rebound effect.

CHAPTER 11: 'PSYCHOSIS' AND HEALING THROUGH RELATIONSHIP

What is defined as schizophrenia and psychosis is typically a state of chronic fear and terror. These individuals have been shattered by trauma. Within them, mental images of past events continue to haunt them. The inner voice (or conscience) which we all possess becomes amplified to a level where visual and auditory hallucinations become present. Grandiose thoughts arise as an attempt to either stave off depression or to escape from the painful reality of a distressing situation and disordered world. Anti-psychotics have been used to diminish the hallucinations and other distressing behaviors, but they have never addressed the reactions of the person and the underlying trauma and factors that has led them to seek a departure from defined reality. Therefore, in collaborating with these individuals, we must meet them in their sense of reality. We must join in respectfully and in a dignified manner, slowly and gently addressing the various disturbances in thought process. We must uncover the hidden traumas and seek to 'be with' the person as they develop new coping mechanisms. It is entirely possible for individuals even in the states of severe mental anguish and distress to recover. And it is indeed possible for this to be accomplished without the addition of toxic drugs. The key is relationship. That is what these individuals are lacking and need. They need to know that there may be exist, if even but one, stable and loving relationship in a world so often filled with pain.

CHAPTER 12: RESULTS OF A STUDY OF THE CAREGIVER'S SKILLS PROGRAM VERSUS MEDICATION AS AN ADJUNCT TO TREATMENT IN A WRAPAROUND PROGRAM

I completed a study of children in wraparound services in Northeastern Pennsylvania. I examined whether a social reinforcement based program designed by Dr. David B. Stein that sets clear target behaviors and emphasizes immediacy and consistency would be more effective than use of psychotropic medications in children diagnosed with disruptive behaviors disorders. The study examined 20 children, 10 receiving medications, and 10 who were not receiving medications and using the Caregiver's Skills Program. The study was over a four month period. The results showed that those receiving the Caregiver's Skills Program had slight improvement over the children receiving medications. In addition, one child on medication was hospitalized because of adverse events from the medications. Therefore, we can assume that the Caregiver's Skills Program is a unique and effective program that does not come with the risks associated with psychotropic medications.

CHAPTER 13: CHILDREN: OUR TREASURE

In today's society we have become so preoccupied with mundane things that we have lost sight of our treasure- our children. In every aspect of society, children are receiving less than what they deserve. Quality time and affection are lacking in families. It is few moments that we take to seek to understand their emotional world and to truly expand our relationship. The education system has failed our children by training them to become good test takers but now endowing them with real skills to succeed and not providing them opportunities to explore their own strengths and inner qualities. It is no wonder that psychiatric diagnoses such as ADHD abound in American society. However, the attention deficit is often not with our children but is rather with the society itself. We have failed to give our children the needed attention. As a result, they react in various ways, sometimes hostily, in order to alert us to our distress. Instead of hearing their distressed voices, we instead resort to drugging them or ordering them into submission to a disordered world. We are looking for externals, particularly drugs, to resolve our problems. But they do not, and in some instances children on anti-depressants have committed suicide and children on stimulants have developed psychosis. Even because a drug is 'FDA approved' does not mean it is ethical and good nor even safe, Let us remember Vioxx as well as many other drugs approved by the FDA later to have catastrophic results for those taking them. I assume psychiatrists can say they 'treated' the problem, as these children are no longer hyperactive, depressed, etc., rather they are dead. None of the so-called disorders can be validated as 'diseases'. There are no quick fixes though many seek them out. It takes hard work to be a parent, hard work to reach our children. But is certainly worth the effort!

CHAPTER 14: THE POWER OF 'JOINING'

In my work with autistic children I have found that there is significant power in 'joining in'. When I wrote the article, "Entering Their Imaginative World", this is what I was referring to by this title. We do not coerce the children to do things as our world sees them, but we seek to enter, understand, and respect the autistic child's world. Through this joining in, such simple things as hopping on one foot with a child who is hopping on one foot, I have seen an extraordinary development of communication and relationship. It was this very process of joining in where I personally saw a 3 year old non-verbal child who frequently engaged in self-stimulatory behaviors move towards communication through reciprocal dialogue, taking my hand and leading me to play with him and a peer, and the building of relationship.

I relate this story from John Clay's book R.D. Laing: A Divided Self. In this quote it describes one of Laing's times where he joined in with a person undergoing serious emotional distress and relationship was forged. Though certainly an unorthodox approach, it worked because it involved joining in and realizing that what may appear to others as seemingly meaningless behavior is meaningful for the individual.

"While still in Chicago, Laing was invited by some doctors to examine a young girl diagnosed as schizophrenic. The girl was locked into a padded cell in a special hospital, and sat there naked. She usually spent the whole day rocking to and fro. The doctors asked Laing for his opinion. What would he do about her? Unexpectedly, Laing stripped off naked himself and entered her cell. There he sat with her, rocking in time to her rhythm. After about twenty minutes she started speaking, something she had not done for several months. The doctors were amazed. 'Did it never occur to you to do that?' Laing commented to them later, with feigned innocence. (pp. 170-171)"

CHAPTER 15: CREATING A SANCTUARY FOR CHILDREN WITH EMOTIONAL CHALLENGES IN OUR SCHOOLS

Children need to be empowered to feel that they are capable of responsible action. Material reinforcement creates dependency, therefore one should utilize social reinforcement. Children need in a voice in the classroom and the ability to use shared energy to solve dilemmas, thus I recommend the use of peer council to deal with problematic behaviors and conflict. Allow the children to seek to resolve dilemmas and there will be amazing results and collaboration rather than an external authority making all decisions. I recommend the use of a 'safe space' and 'dramatized grievances'. The child should be allowed to know that there is a place they can go when they feel distressed or overwhelmed and they can share their distress without judgment. A therapeutic minded adult should be calm and patient and seek to hear what is being said through the behavior, this may be the only way the child can communicate their distress. I also recommend use of listening partnerships, allow the children to have time to discuss interests, positives, concerns, areas of distress by allowing each a 3 minute time for dialogue during the day without interruption. Meditation exercises can help children train their minds to deal with conflict and distraction in a more responsible manner.

Encourage independent exploration, hands on projects and other projects where a child can share their own interests and feel a sense of pride are crucial.

The zeal for learning should be instilled, learning should not be seen merely a task but something desirable.

CHAPTER 16 RELIGION AND SPIRITUALITY

A colleague recently asked me if I felt religion was a positive or destructive force, arguing how that many conflicts in the world appear to have a religious base. Religion is an established set of beliefs often accompanied with rituals and an ethical plan that codifies how we should look at the world around us and our interactions with others. In some sense, every person, even those without a belief in a deity, could be said to hold to a 'religion'. The issue is how organized and dogmatic this particular religion may be. Whether religion is a positive or destructive force is dependent on the means by which we choose to use it. If this defining of beliefs seeks to provide meaning and guides us in fulfilling relationships, it can be said to be positive. If it is used to merely control the conduct of others or force them to alter their nature, it can become oppressive. Religion then can be a means when there is equanimity and shared values and meaning, a positive force which does help to guide human interactions. If the religion is based merely on fear and intimidation to guide human interactions, it becomes destructive. I have often argued that in today's society, psychiatry has become a religion and is now the 'standard' by which to codify human nature and behavior and to intimidate those who would be deviants. Those who are considered 'mentally ill' are the heretics of this religion. Then there is also those who seek to differentiate between religion and spirituality. "I am not religious, but spiritual." Good heavens, what does this mean? It often appears that individuals try to seek something outside of nature to find meaning. But what exactly is wrong with nature? Why it is that man seeks to somehow 'improve' upon nature? Why can we not accept the majesty and also the mystery of the natural world around us? It often appears that those that cannot find delight in the nature of children want to label them, drug them, or somehow alter their nature to fit into a defined way of how they feel children should behave. But this does not only apply to children. Therefore, when we define 'spiritual', we must carefully understand this term? Are we referring to our own nature and its potential or are we directing this search to something outside of us, something that must be altered? Spiritual should refer to the mind. The mind is not the brain; it is our conscience, our experiences, our nature, and ultimately who we are. To use the term 'spiritual' as recognizing our own potential for benevolence, then this is a positive use of this term.

Lightning Source UK Ltd.
Milton Keynes UK
UKHW012303090223
416755UK00001B/110